The Cult of Pyth

THE CULT OF PYTHAGORAS

Math and Myths

ALBERTO A. MARTÍNEZ

University of Pittsburgh Press

Published by the University of Pittsburgh Press, Pittsburgh, Pa., 15260
Manufactured in the United States of America
Printed on acid-free paper
First paperback edition 2013
10 9 8 7 6 5 4 3 2 1

ISBN 13: 978-08229-6270-0
ISBN 10: 0-8229-6270-5

Library of Congress Cataloging-in-Publication Data

Martinez, Alberto A.
The cult of Pythagoras : math and myths / Alberto A. Martinez.
    p.   cm.
Includes bibliographical references and index.
ISBN 978-0-8229-4418-8 (hardcover : alk. paper)
1. Mathematics—History. 2. Geometry—History. I. Title.
QA21.M335 2012
910.938—dc23   2012022366>

From of old it has been the custom,
and not only in our time,
for vice to make war on virtue.
Thus Pythagoras, with three hundred others,
was burned to death.

—*Cohort ad Græcos*, ca. 150 CE

CONTENTS

LIST OF MYTHS AND APPARENT MYTHS

ACKNOWLEDGMENTS

I warmly thank the following friends and colleagues for kindly sharing thoughtful comments and helpful suggestions in various aspects of this project: Eric Almaraz, Ronald Anderson, Casey Baker, Nicole Banacka, Frank Benn, Susan Boettcher, Jochen Büttner, Richard Chantlos, Alexis R. Conn, Robert Crease, Chandler Davis, Jason Edwards, Rodrigo Fernós, Paige Fincher, Richard Fitzpatrick, Paul Forman, Allan Franklin, Craig G. Fraser, Jennifer Rose Fredson, Jeremy Gray, Roger P. Hart, Jess Haugh, Vanessa Hayes, Reuben Hersh, Judy Hogan, Bruce J. Hunt, Daphne Ireland, Alexander R. Jones, Akihiro Kanamori, Daniel Kennefick, Stephen J. Lundy, Michael Marder, Ronald Martínez Cuevas, Rubén Martínez, Anahid Mikirditsian, Lillian Montalvo Conde, Michelle Morar, Bruce Palka, Thomas Pangle, David Lindsay Roberts, César Rodríguez-Rosario, Lynn Robitaille, Heather Rothman, Jenna Saldaña, Claudia Salzberg, Alan E. Shapiro, Ali Shreif, Christopher Smeenk, Angela Smith, John Stachel, Sarah Steinbock-Pratt, Paige L. Sartin, Ariel Taylor, Eve Tulbert, Heather Turner, Kirsten van der Veen, Scott Walter, Lauren Ware, James A. Wilson Jr., Sarah D. Wilson, and Detou Yuan.

Next, I thank the helpful librarians of the following universities: the University of Texas at Austin; Harvard University; the California Institute of Technology; the Einstein Archives; the Dibner Library of History of Science and Technology at the National Museum of American History, Smithsonian; the Burndy Library (when it was located at the Dibner Institute for the

History of Science and Technology, at MIT); the Landesbibliothek Bern; the Archives and Special Collections of the Bibliothek of the Eidgenössische Technische Hochschule in Zurich; and the library of the Poincaré Archives of the Université Nancy 2. Next, I thank the following editors for their kind advice, support, and suggestions: Ann Downer-Hazell, Elizabeth Knoll, Michael Fisher. Moreover, I also thank editors Phyllis Cohen, Vickie Kearn, and Maria Guarnaschelli. Finally, I also thank eleven anonymous reviewers for their various comments and suggestions.

I am especially thankful to the University of Pittsburgh Press for kindly adopting my evolving manuscript. Its acquisitions editor, Beth Davis, enthusiastically supported the project and steadily contributed to its fruition and shape. Beth left the press in mid-2011, but subsequently Peter Kracht and his assistant, Amberle Sherman, have diligently supported the project. I also appreciate the fine, careful work carried out by Carol Sickman-Garner, in copyediting my manuscript. I thank production director Ann Walston for the beautiful design and typesetting, and I also thank managing editor Alex Wolfe for overseeing the copyediting process. The success of most books also depends greatly on the support provided by the marketing crew, and hence I warmly thank Lowell Britson, David Baumann, Maria Sticco, and the staff at the University of Pittsburgh Press for their gracious support.

A NOTE ON TRANSLATIONS

In my efforts to emphasize the importance of using primary sources, I carried out many of the translations from sources in French, German, and Italian. Latin and Greek are beyond my level of competence, so in many cases I used published translations for ancient languages. Nevertheless, wherever word choices seemed important (as in the ambiguous ancient sources that refer to Pythagoras in relation to mathematics), I did painstakingly translate ancient passages from Latin and Greek. Afterward, I received some helpful assistance from Van Herd, who helped me to clarify some questions of word choice and polish the grammar. Having done this, I finally had a fair grasp of the overall meaning and actual words involved in each such passage. Still, I then proceeded to solicit the help of Erik Delgado, trained in classics, to carefully check most of my translations from Latin and Greek and essentially to retranslate each passage in an utterly literal way. I then edited his translations slightly for word choice, in due regard to the original sources. In the end, I can guarantee that our translations are at least more literal than most of the well-known English translations of ancient passages about Pythagoras. At the same time, I apologize if some of our expressions seem coarse: my goal has been fidelity, not smoothness, so we have not engaged in "filling in the blanks," hiding ambiguities, or using later accounts to set the meaning of earlier sources.

I did not include untranslated passages in the book; however, I think that one example might help to illustrate why translations from primary sources are important. Consider the words of Athenaeus, from around 200 CE, in the original Greek:

καί θῦσαί φησιν αὐτὸν ἑκατόμβην ἐπὶ τῷ εὑρηκέναι ὅτι τριγώνου ὀρθογωνίου ἡ τὴν ὀρθὴν γωνίαν ὑποτείνουσα τείνουσα ἴσον δύναται ταῖς περιεχούσαις·

ἡνίκα Πυθαγόρης τὸ περικλεὲς εὕρετο γράμμα,
κλεινὸς ἐφ᾽ ᾧ κλεινὴν ἤγαγε βουθυσίην.

One popular translation of this passage was published by Charles D. Yonge in 1854:

he even sacrificed a hecatomb when he found out that in a right-angled triangle, the square of the side subtending the right angle is equal to the squares of the two sides containing it—

When the illustrious Pythagoras,
Discovered that renowned problem which
He celebrated with a hecatomb.

By contrast, compare Yonge's translation to my very literal translation:

he even sacrificially burned a hecatomb upon finding out that in a right-angled triangle the hypotenuse subtending the right angle equals in power its peripherals:

When Pythagoras found the revered inscription
the celebrity brought for it a celebrated ox-sacrifice.

Yonge's translation adds the words "square," "squares," "side," and "two sides." Such clarifications are fine, but these words are absent in the original. Yonge also construed the ambiguous βουθυσίην (sacrifice of one ox or more) as a hundred, a hecatomb, and he changed the ambiguous γράμμα for a "problem" to be solved. It sounds plausible, but that is not what the original says.* The present book asks: how does history change when we subtract the many small exaggerations and interpolations that writers have added for over two thousand years?

* The word γράμμα has also been interpreted as diagram, thus Plutarch (ca. 100 CE) quoted the same epigram and added the word διαγράμματι. However, following the principle that later commentaries should not be used to set the meaning of earlier sources, I do not interpret γράμμα necessarily as a geometrical diagram. In Euclid's *Elements*, γραμμή means line, παραλληλόγραμμον appears often, and εὐθύγραμμα means straight-line figure. But γράμμα by itself is not used, instead, other words are used to mean geometrical diagram or figures: καταγραφή, σχημάτων (e.g., bk. 3 prop. 33, bk. 4 prop. 5). Moreover, γράμμα had other common meanings: letter, inscription, portrait. Consider the following examples. Herodotus, *Histories* (ca. 430 BCE) 1.139: τελευτῶσι πάντα ἐς τώυτὸ γράμμα (all end in the same letter); *Histories* 1.148: ἐς τώυτὸ γράμμα τελευτῶσι (end in the same letter). Xenophon, *Memorabilia* (ca. 375? BCE) 4.2.24: πρὸς τῷ ναῷ που γεγραμμένον τὸ γνῶθι σαυτόν; ἔγωγε. πότερον οὖν οὐδέν σοι τοῦ γράμματος ἐμέλησεν ἢ προσέσχες (did you notice somewhere on the temple the inscription "Know thyself"? And did you pay no heed to the inscription). Plato, *Phaedrus* (ca. 370 BCE), 229e, 230a: "τὸ Δελφικὸν γράμμα" (as the Delphic inscription). Plato, *Republic* (ca. 375 BCE), 472d: ἀγαθὸν ζωγράφον εἶναι ὃς ἂν γράψας παράδειγμα οἷον ἂν εἴη ὁ κάλλιστος ἄνθρωπος καὶ πάντα εἰς τὸ γράμμα ἱκανῶς ἀποδοὺς (a good painter, who after drawing a design of the most beautiful man and omitting nothing required for the perfection of the portrait). Theocritus, *Idylls*, 15.81 (ca. 275 BCE or much later): ζωγράφοι τἀκριβέα γράμματ᾽ ἔγραψαν (such work, what the painter painted). *Anthologia Palatina*, 6.352 (allegedly by Erinna; unclear origins, 600 BCE to 600 CE): Ἐξ ἁπαλᾶν χειρῶν τάδε γράμματα (delicate hands made this portrait). Therefore, regarding the poetic words about Pythagoras: did γράμμα mean letter, portrait, inscription, image, or diagram? I chose the word "inscription" partly because it captures the ambiguity, rather than disguising it: an inscription can be a sentence or it can be a diagram; it can be something handwritten or it can be an etching.

INTRODUCTION

The international bestseller *The Secret* claims that Pythagoras knew the secret to happiness, the powerful law of attraction: that you can get what you want by thinking about it. Less recently, in one of the most popular science books ever, Carl Sagan noted that on the island of Samos local tradition says that their native son Pythagoras was "the first person in the history of the world to deduce that the Earth is a sphere." Earlier, the mystic architect John Wood, having studied megalithic ruins such as Stonehenge, concluded that they were "a Model of the Pythagorean World," that such stones were set by druids who followed the main priest of Pythagoras.[1]

But none of this is true: there's no evidence for it. This book is about the evolution of myths in the history of mathematics. It's also about invention, about how writers create imaginary histories of their favorite topics. The case of Pythagoras shows a common mismatch between speculations and evidence in history. For the public at large his name is the most famous name in mathematics: millions of people who have never heard about Euler, Gauss, or Galois have nevertheless heard about Pythagoras. But strangely, there's hardly any evidence that he contributed anything to mathematics. How did this leader of a small and secretive religious cult become world famous in mathematics? A series of great achievements were attributed to him for two thousand years.

Myths are contagious fiction. They're charming stories that arise repeatedly and spread. I'll discuss various popular myths: that Pythagoras proved the hypotenuse theorem; that he believed that the world is made of numbers;

that his cult murdered Hippasus for discovering irrational numbers; that Euler was confused about multiplying imaginary numbers; that the bright boy Gauss instantly added the integers from 1 to 100; that Galois created group theory on the night before he died in a pistol duel; that the golden ratio is the most beautiful number, loved by the ancients; and more. Some stories are partly true, others are entirely false, but they all show the power of invention in history. These stories grow in the writings of mathematicians, educators, famous scientists, and in Disney and *The Da Vinci Code*. By reflex, writers and teachers stretch the meaning of what they read. They add meaning, supposedly hidden but essential.

This book is also about invention in a very positive sense. People usually view mathematical breakthroughs as a series of discoveries. They assume that numbers and rules existed timelessly, before being discovered. But why? Consider the history of irrational numbers, zero, negative numbers, imaginary numbers, quaternions, infinity, infinitesimals, triangles, and circles. Was there ever invention in these concepts? William Rowan Hamilton and Georg Cantor believed they had discovered numbers that had existed forever. Gottfried Leibniz and Abraham Robinson believed instead that the new numbers they described were essentially fictions, invented.

Recently, mathematician Paul Lockhart has argued that mathematics is a creative art and that teachers should teach its important aspects that are usually omitted: invention, history, and philosophy. Most students learn the principles of math dogmatically, like articles of faith. Lockhart complains: "Students are taught to view mathematics as a set of procedures, akin to religious rites, which are eternal and set in stone. The holy tablets, or Math Books, are handed out, and the students learn to address the Church elders as 'they.' (As in 'What do they want here? Do they want me to divide?')"[2] Incidentally, a mathematics editor told me what her young son once said to a question about his family's religion: "My dad is Christian, my mom is Mathematician."

Many people view mathematics as essentially different from other fields, uniform, as if mathematicians always agree. It's a well-kept secret: actually, mathematicians sometimes disagree. They've disagreed about what is possible and impossible, they've disagreed about what counts as a proof, they've even disagreed about the results of certain operations, and most often, they disagree about the meaning of mathematical concepts. Why keep this interesting secret from students? Some mathematicians might deny that this secret exists. But ask any student in high school to state even *one* example of mathematicians ever disagreeing about the result of an operation. Most students don't know any such example, because their teachers haven't taught one.

In 2001, the magazine *Physics World* ran a poll on the philosophical views of physicists. Among various questions, about the reality of electrons, genes, atoms, emotions, and light waves, the survey also asked about beliefs regarding numbers. Some respondents wrote "not sure" about some questions; a few just didn't reply. But many physicists did submit their answers, and a total of 534 replies were received.[3] For example, the poll produced the answers in table 1, with each percentage stating how many respondents chose that option.

Table 1

	Real	*Not real*
The Earth	93%	3%
Stones	93%	3%
Genes	83%	8%
Electrons	84%	9%
Light waves	68%	20%
Real numbers	66%	26%
Imaginary numbers	43%	44%

The author of the poll, philosopher and historian of science Robert Crease, acknowledged that the word "real" has various meanings to different people. But generally, people view something as real if they think that it exists in the world independent of human perceptions and thoughts. It's interesting that out of hundreds of respondents, most of them physicists, less than half of them regard imaginary numbers as real.

It would be revealing to carry out the same kind of poll with mathematicians. Some might think that the question of whether real or imaginary numbers are "real" is meaningless. But I disagree: I think that certain aspects of mathematics are made by our imaginations, and others describe relations that exist independently.

At the University of Texas at Austin, I teach a course that is required for majors in mathematics or the sciences who want to become teachers, through the UTeach program. Every semester I give students a survey asking a few questions about their views on mathematics. Some of those questions have never been raised by their teachers. The survey asks:

1. Triangles existed before humans and will continue to exist forever. True or false?

2. Circles are _____ real than apples. (Less? equally? more?)

Students kindly disregard the grammatical incoherence in the second query and just answer the questions. From 2005 to 2010, undergraduate students in thirteen groups answered surveys including these questions, the results of which can be seen below. Out of 245 majors in mathematics and the sciences over those five years, 77 percent of the students wrote that triangles existed before humans and will continue to exist forever. Almost 22 percent disagreed, and only 3 students chose not to reply and wrote instead "maybe," "neither," and "no idea." The answer given by the majority of students matches the views of many prominent mathematicians for centuries. It helps to explain why some individuals labored for decades on certain problems and why they spoke about mathematics in religious terms. However, if it's false that triangles are eternal, if perhaps triangles are concepts invented by humans, then we should discuss the value of fiction in the elements of mathematics.

Triangles existed before humans and will continue to exist forever. True or False?

Likewise, five years of replies to a similar question about circles are conveyed in the next figure. Out of 245 students, the great majority claimed that circles and apples are equally real, while roughly 1 out of every 4 said that circles are less real, and only 2 students gave no answer. Eleven students said that circles are *more* real than apples, which resembles Plato's ancient philosophy: the notion that mathematics describes a realm of changeless, eternal ideas that are more real than material things, which take shape but later dissolve.

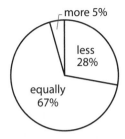

Instead of the notion that mathematics is about eternal immaterial objects, I'll argue that it's more about rules for manipulating concepts. I will argue that a mathematician's productivity and creativity are directly motivated by that person's views on the reality of mathematical objects. Yet a common attitude is that philosophy and history are nearly irrelevant for mathematics. When physicists responded to Robert Crease's survey on the reality of things, a few of them pointedly submitted the survey questions entirely blank, and one of them complained: "In the last 40 years, philosophers still have not found a way to ask a physicist a 'real' question." Similarly, some students entering my course openly express skepticism that history or philosophy can be of any use to them. Every year I meet some students who think that science has no connection to mathematics. They actually say it: "I don't see why we have to study science—I'm a *math* major." I've heard such words several times: it's not a random comment, but it makes sense given the old popularity of the so-called Platonist philosophy, despite a recent downturn.

History shows that different rules have been proposed and used for operating on numbers and geometry. Mathematicians have sometimes disagreed about dividing by zero, multiplying and dividing imaginary numbers, handling infinitesimals, and so on. While discussing seemingly impossible operations, I've tried to make every chapter very understandable, avoiding jargon and needless complexity. As rightly noted by mathematician John Allen Paulos, "It is almost always possible to present an intellectually honest and engaging account of any field, using a minimum of technical apparatus. This is seldom done, however, since most priesthoods (mathematicians included) are inclined to hide behind a wall of mystery and to commune only with their fellow priests."[4]

But debunking is a delicate business. We have to appreciate that there are certain delights in conjectures. A friend has a tattoo of what a classics professor once seriously told her: *History is fiction.* Likewise, another friend told me what she always liked about history, what was fun about it, "that you get to fill in the blanks." There are also delights in confusions. A friend of a friend sent me an anecdote that nicely shows how easily people confuse moments of "discovery." In 2008 he was teaching a class at a language institute, one of the best in Beirut. They were discussing "genius," and he made a reference to Einstein. But surprisingly, the students were not sure who Einstein was, and then one of them, a lawyer, seriously said: "Sir, isn't he the man who was taking a bath, when an apple fell on his head and he discovered the word *eureka*?" He discovered the word! The teacher wrote to me: "I'm sure her abilities to express herself in English played a part, but her words will nonetheless be etched in my memory forever."

We will analyze myths and invention in mathematics. I'll criticize how writers have invented history, but more important, I will pinpoint and praise invention in the growth of mathematics itself. In my research, surprisingly, I repeatedly found Pythagoras, in ways that differ greatly from what we might expect. In this book, the legendary Pythagoras will march prominently on center stage at first, but later he will recede to lurk in the background. He will emerge as the patron saint of the urge to pretend to know the past. Now let's take a look at the creeping shadow of the elusive Pythagoras.

The Cult of Pythagoras

TRIANGLE SACRIFICE TO THE GODS

Legends say that in ancient times a secretive cult of vegetarians was led by a man who had a strange birthmark on his thigh and who taught that we should not eat beans. He believed that when a person dies, the soul can be reborn in another body, even as an animal. So he said that we should not eat animals because they might be our dead relatives or friends. And he said that he had been born five times, even before the Trojan War. And when he died his fifth death, his followers later said that he was reborn again. But why should we not eat *beans?* His reason: that if a bean is moistened and placed in a pitcher and buried, when we dig it up days later we will find a disturbingly familiar form growing: the head of a human child.[1]

We might think that none of this is true. But still, you've heard of the alleged fifth incarnation of this cult leader. He was Pythagoras, born sometime around 570 BCE and dying roughly seven decades later.[2] People know him as the mathematician who discovered the Pythagorean theorem: that the squares on two sides of a right triangle add up to the square on its largest side.

But wait. There is no evidence that he discovered that. It was already well known to the Hindus and the Chinese.[3] And the Babylonians knew it more than a thousand years before Pythagoras was born on the island of Samos in the Aegean Sea.

Some people say instead that Pythagoras was the first to *prove* that the theorem is true for all triangles. But again, is there any evidence that he did that? Lacking evidence that he really did what gives him worldwide fame, some teachers and historians guess that Pythagoras *may have* been the first to prove the theorem.

May have been the first? Maybe he was the fourth? Or maybe he just never did that? Maybe he was reborn seventeen times?

Like other teachers, I too used to tell my students widespread stories about Pythagoras. But eventually I became uncomfortable in not knowing the roots of those stories. What if we confront such uncertainties?

If Pythagoras wrote anything, none of his writings seems to have survived. What remains was written long after his death. We don't trash it all because some of the various fragments are mutually consistent and because some reliable commentators apparently wrote some passages. Ancient sources don't say that Pythagoras proved the hypotenuse theorem, but they do say that he didn't eat beans and that he believed that souls are reborn.

They also tell other stories. For example, that Pythagoras never laughed.[4] That he infallibly predicted earthquakes, storms, and plagues. That he said that earthquakes are conventions of the dead.[5] Also, that "there was such persuasion and charm in his words that every day almost the entire city turned to him, as to a god present among them, and all men ran in crowds to hear him."[6] And, that when he and his associates once crossed the river Nessus, Pythagoras spoke to the river, and it loudly replied: "Hail, Pythagoras!"[7] One ancient poem says that Pythagoras was the son of the god Apollo, who visited his mother: "Pythagoras, whom Pythias bore for Apollo, dear to Zeus, she who was the loveliest of the Samians."[8]

But is any of this true? What can we believe about a mysterious man who lived ages ago? And why does it matter? It matters because by trying to replace legends with history we exercise critical thinking. Some people don't want to discard a familiar simple story. But by seeking evidence we learn to pinpoint falsehoods; so Albert Einstein remarked: "Whoever is careless with the truth in small matters cannot be trusted in important affairs."[9]

We tend to care about long-dead people to the extent that we find ourselves reflected in them. Math teachers care about Pythagoras mainly because they construe him as an ancient role model, a hero who got something right, a genius who recognized the importance of mathematics. Likewise, vegetarians admire Pythagoras because he said that we should not eat animals. Musicians admire him because he allegedly discovered numerical ratios in the lengths of strings that make harmonies.

But was he even a mathematician? His admirers linked numbers and religion. Did their religious beliefs affect our views on mathematics?

Pythagoras matters because his name is *the* most common historical element in schoolbooks on mathematics. To say anything fair about him, we must enter a maze of hearsay. Consider first the most common story: that he discovered the Pythagorean theorem.

A mystic pagan theologian, Proclus, wrote an influential account around 460 CE. Proclus briefly claimed that Pythagoras based a liberal form of education on geometry, that he investigated its theorems in an immaterial way, and that he initiated the study of irrational magnitudes and the five regular solids. Also, Proclus explained that the sum of squares on two sides of a right triangle equals the square on its hypotenuse, and he briefly wrote, "It is possible to find men listening to men wishing to inquire into ancient things, relating this theorem to Pythagoras, and calling him 'ox-sacrificer' for his finding."[10] It's unclear whether this means that Pythagoras himself made the discovery or that he just was impressed when he heard about it. And it's unclear whether Proclus even agreed with "those wishing to inquire into ancient things." Another problem with his words is that he wrote them nearly *a thousand years after* Pythagoras. A story can change in a day, so how much might it change in *a thousand years?*

On what did Proclus base his comments? He had access to a copy of an ancient history of geometry, first written by Eudemus around 330 BCE, now lost. But Proclus did not specify Eudemus, as he did for other topics. Some of Proclus's comments on Pythagoras were copied (in some places word for word) from recent works written around 300 CE, mainly by the Syrian philosopher Iamblichus.[11] He based his writings partly on old accounts, but Iamblichus also included many exaggerations to glorify Pythagoras. Allegedly, Pythagoras invented political education, overthrew despotic regimes, freed cities from slavery, and entirely abolished discord and differences of opinion in and among all cities in Italy and Sicily, for many generations.[12] Furthermore, Pythagoras supposedly founded the science of harmonics, and Iamblichus also ascribed to him the theory of astronomy that was actually developed by Ptolemy long after Pythagoras had died.[13] He also claimed that Pythagoras coined the word *philosophy* and began its discipline.[14] And that Pythagoras rejected foods that cause gas; that he spoke to a bull, convincing the bull not to eat beans; and that he used numbers instead of animal entrails to divine the future.[15] Above all, Pythagoras was a superhuman miracle worker sent from the domain of the god Apollo to enlighten humans to live properly. Iamblichus claimed that Pythagoras was "the most handsome and godlike of those ever recorded in history."[16]

Iamblichus was not a historian. He embellished Pythagoras to make him seem a gifted and inspirational figure to tell us how to live well. As for the story about Pythagoras and the triangle, we don't know where Proclus got it, but it appeared earlier. Iamblichus's teacher, the philosopher Porphyry, wrote a shorter biography of Pythagoras. He too praised Pythagoras as being superhuman. Porphyry celebrated Pythagoras as an ancient moral figure much preferable to Jesus Christ. Porphyry, in addition, wrote fifteen books titled *Against the Christians* that were banned by the Roman emperors Constantine and Theodosius: his books were burned, destroyed.[17]

In his biography, Porphyry said that Pythagoras had a golden thigh: evidence that he was divine, related to the god Apollo. He said that Pythagoras predicted earthquakes and stopped violent winds, hail, and storms over rivers and seas. And that Pythagoras shot an arrow that carried his priest to practically walk on air. And he commented: "Of Pythagoras many other more wonderful and divine things are persistently and unanimously related, so that we have no hesitation in saying never was more attributed to any man, nor was any more eminent."[18] Porphyry told about the hypotenuse and the ox: "Sacrificing to the gods, he was without offense, propitiating the gods with barley and cake and frankincense and myrrh, but least of all with living things, except with fowl and the most tender parts of piglets. He once sacrificed an ox made of flour, so the more precise accounts say, having found that in the rectangle the hypotenuse equals in power its peripherals."[19] So maybe Pythagoras did not kill an ox?

There are earlier versions of the story. At around 225 CE, Diogenes Laertius wrote about Pythagoras, without worshiping him. He wrote: "Apollodorus the logician says that he sacrificed a hecatomb upon finding out that in the right-angled triangle the hypotenuse side equals in power its peripherals. And, in sum, the epigram thus conveys: 'When Pythagoras found the revered inscription, that man brought for it a celebrated ox-sacrifice.'"[20]

By this point one might think, "Should we read these quotations? Don't they all say the same thing?" That's the point: they don't, so instead of proceeding as usual, writing something to the effect of "believe me, this is how it went," giving some generalization, we should check the evidence itself, add it up, to see where it takes us. Writers often misrepresent sources by paraphrasing—for example, by writing that Diogenes credited Pythagoras with "the proof" of the hypotenuse theorem.[21] But it's not true! As I tried to figure out the stories about Pythagoras, I was very surprised, sometimes shocked, to discover how often writers just invent words and ideas that are not in the original texts. Translations usually include many words that simply do not exist in the originals. So to analyze these old stories, we're using new translations from Latin and Greek, which are more accurate than any others. So what did they really say?

A hecatomb means a hundred oxen, so allegedly Pythagoras not only discovered a relation about triangles but celebrated by bloodily killing many oxen.[22] Diogenes's words are vague: was the famous "inscription" a letter, sentence, drawing, or diagram? Was it even geometrical? Diogenes alluded to an Apollodorus, of whom we have no surviving texts, but a couple of earlier writers mentioned him. Around 200 CE, Athenaeus wrote:

> Apollodorus the arithmetician says he even sacrificially burned a hecatomb upon finding out that in a right-angled triangle the hypotenuse subtending the right angle equals in power its peripherals:
>
>> When Pythagoras found the revered inscription,
>> the celebrity brought for it a celebrated ox-sacrifice.[23]

And an earlier account was written by Plutarch, a Greek priest and essayist. At around 100 CE, he wrote:

> And Pythagoras sacrificially burned an ox for his diagram, as Apollodorus says:
>
>> When Pythagoras found the revered inscription,
>> the celebrity brought for that a radiant ox-sacrifice.
>
> whether about how the hypotenuse equals in power its peripherals, or the problem of the area of the parabola.[24]

First, this passage was written about *seven hundred years* after Pythagoras had died. Second, the last sentence suggests that Plutarch was unsure whether Pythagoras had sacrificed an ox because of one problem or another. Third, what inscription? The brief words are vague; we don't know the poem from which they came. We don't know who this Apollodorus was, when he lived, or whether there is any reason to take his one sentence as echoing any historical events.[25]

Another version of the story appears in *De Architectura*, by Vitruvius. He completed his books around 15 BCE, but we don't know which portions are original and what was added by later writers. The earliest surviving manuscript copy, from around 815 CE, includes these lines:

Likewise Pythagoras showed the set-square, found without the fabrications of a craftsman, and that which workmen, building a set-square, can scarcely produce accurately, which having been corrected by calculations and procedures is explained because of his precepts. For if three rods are given, one of which is 3 feet, the second 4 feet, the third 5 feet, and these rods, positioned among themselves, should touch each other by their extremities, making the shape of a triangle, they will form the shape of the corrected set-square. And if on each of those rods' lengths are drawn single squares with equal sides, that of side 3 will have an area of 9; the 4, 16; that of 5, 25.

Thus, the footage area of two squares, on the side 3 feet long, plus that of 4, are numerically as large as made by one drawn on the 5. When Pythagoras had found this, not doubting that the Muses had helped him in this invention, with the greatest gratitude he allegedly sacrificed victims to them. The same calculation, moreover, is useful in many things and measurements, even in buildings for the construction of stairs, in order to expedite their having measured divisions of steps.[26]

Did Vitruvius himself write this? In any case, the passage does not describe a rule for all right triangles; it refers only to the 3, 4, 5, triangle.

The earliest extant account of Pythagoras and the ox is from around 45 BCE. The Roman statesman Cicero wrote: "Nor did anyone ever pledge a tithe to Hercules, if to become a sage—although Pythagoras, upon finding something new in geometry, is said to have immolated an ox for the Muses; but I do not believe it, since he did not even want to immolate an animal for Apollo at Delos, to not sprinkle the altar with blood. But back to the issue, this is the reckoning of all mortals, that fortune is sought from God, but wisdom must be taken from oneself."[27] This is the earliest known account. And Cicero nowhere specified that it pertained to what became known as the Pythagorean theorem. So, some 450 years after Pythagoras had died, the story of him killing an ox had arisen, and the writer who relays it dismissed it as false.

Cicero said that Pythagoras refused to kill for the gods. Is there more evidence for that? Diogenes (ca. 225 CE) noted that some writers claimed that Pythagoras sometimes sacrificed animals and that one of his followers, Aristoxenus (who never met Pythagoras), said that Pythagoras "permitted the eating of all other animals, and abstained only from oxen used in agriculture, and from rams." But Diogenes also wrote that

> in reality he prohibited the eating of animals because he wished to train and accustom men to simplicity of life; so that all their food should be easily procurable, as it would be, if they ate only such things as required no fire to cook them, and if they drank plain water; for from this diet they would derive health of body and acuteness of intellect. The only altar at which he worshipped was that of Apollo, the Father, at Delos, which is at the back of the altar of Caratinus, because wheat and barley, and cheesecakes are the only offerings laid upon it, as it is not dressed by fire; and no victim is ever slain there, as Aristotle tells us, in his Constitution of the Delians.[28]

Likewise, Iamblichus claimed that Pythagoras neither ate nor sacrificed animals and that he required his closest followers to abstain as well, while letting ordinary people eat some animals, people "whose life was not entirely purified, philosophic and sacred." And Porphyry too noted that Pythagoras did not kill or eat animals. He quoted from an ancient work by Eudoxus, who wrote about one hundred years after Pythagoras had died that "Pythagoras used the greatest purity, and was shocked at all bloodshedding and killing; that he not only abstained from animal food, but never in any way approached butchers or hunters."[29] At around 8 CE, the Roman poet Ovid claimed that "he was the first man to forbid the use of any animal's flesh as human food."[30] At around 50 BCE, Diodorus of Sicily wrote that "Pythagoras believed in the transmigration of souls and considered the eating of flesh as an abominable thing, saying that the souls of all living creatures pass after death into other living creatures."[31]

Other writers also said that Pythagoras did not eat animals. But the earliest sources don't address the point directly.[32] Various sources give apparently conflicting accounts of whether the Pythagoreans ate meat. The conflict might be resolved by distinguishing among the practices of Pythagoras, those of his inner circle of disciples, and his prescriptions for people in general.[33] Finally, consider words that apparently were voiced by Xenophanes, who lived at the same time as Pythagoras and reported: "And they say that once, passing by a puppy being beaten, he took pity, and said this remark: 'Stop, do not beat him, since this is the soul of a dear friend, which I recognized hearing him yelp!'"[34] These words resonate with the claims that Pythagoras would not sacrifice animals and that he believed in the transmigration of souls.

Returning to the story of the triangle and the ox, the trail disappears with Cicero, at around 45 BCE. Nowadays, most writers still credit the hypotenuse theorem to Pythagoras, even on apparently historical grounds: "on the authority of several Greek and Latin authors, including Plutarch and Cicero, who wrote half a millennium after Pythagoras."[35] But the matter is not so clear, as Cicero did not specify the theorem in question, just "something new in geometry"—and again, Cicero doubted the story. And what was the reliability of Plutarch? Some writers quote him as a reputable bearded source, but they do not mention other comments Plutarch wrote: that Socrates "received philosophy from Pythagoras and Empedocles, full of dreams, fables, superstitions and perfect raving," before Socrates tried to improve wisdom.[36] That Pythagoras engaged in false divination and controlled his disciples' curiosity by restricting them from speaking for five years, whereas, "they say, Pythagoras one time ranted a friend of his so terribly before company, that the poor young man went and hanged himself."[37]

So what evidence is there, in the first four hundred years after his death, that Pythagoras found, proved, or even celebrated *any* geometrical theorem? None. We have extant writings from several mathematicians, philosophers, and historians from that time. But none of them claims any such thing.

For example, in the extant treatise by the geometer Apollonius of Perga (who died in 195 BCE), there is no mention of Pythagoras having contributed anything to geometry. Also, at around 225 BCE, the great mathematician Archimedes wrote several comments about the history of geometry, but he did not claim that Pythagoras had contributed anything to it. Likewise, at roughly 250 BCE, Euclid compiled most geometrical knowledge into thirteen books, *The Elements*.[38] This masterpiece discusses the hypotenuse theorem and gives two proofs but does not mention Pythagoras or anyone. Also, while Aristotle wrote about the Pythagoreans and also much about mathematicians, he seems to have reported no such thing. No direct references exist by Aristotle attributing anything in mathematics to Pythagoras. And earlier, Plato, in his various historically grounded dialogues, where he discussed mathematics, seems to have never said that Pythagoras found or celebrated any geometric theorem at all. Instead, Plato elsewhere just characterized Pythagoras as the founder of a way of life, loved by his followers for his wisdom.[39]

So the story that Pythagoras found, proved, or celebrated the hypotenuse theorem dissolves into nothing. Legends are pleasant: we imagine Pythagoras as a brilliant geometer who killed a hundred oxen in a sacrificial bloodbath. But if we doubt *half* of the story—that he killed oxen—then why should we believe the other half, about some discovery?

How can we replace legends? What can teachers fairly say about the origins of the hypotenuse theorem? We can say:

> Several ancient clay tablets show that the Babylonians of the Old Empire knew how to compute the sides of right-angled triangles, between 1900 and 1600 BCE. And early geometrical proofs of the hypotenuse theorem are found in *The Elements,* completed roughly around 250 BCE.

Note that this stops short of attributing discovery or originality to Euclid, as that too is uncertain. But at least we have said something definite and fair. As for Euclid, we should discuss him; one of my students told me that she had *never* heard of Euclid before her senior year in college, in my course, while she had often heard of Pythagoras. In schoolbooks, the most famous ancient geometer is often eclipsed by the religious leader who oddly shines as the author of geometric proof.

Stories are spread by prominent voices. Here is a typical contagious

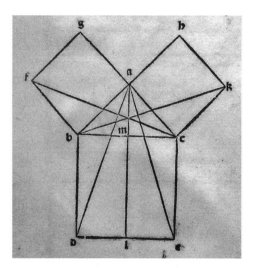

Figure 1.1. Diagram of a proof of the hypotenuse theorem, from Euclid's *Elements,* in a Latin edition of 1491.

example of the claim that Pythagoras introduced proof into mathematics. In the longtime bestselling book *Men of Mathematics,* first published in 1937, the mathematician Eric Temple Bell claimed: "Pythagoras then imported *proof* into mathematics. This is his greatest achievement. Before him geometry had been largely a collection of rules of thumb empirically arrived at without any clear indication of the mutual connections of the rules, and without the slightest suspicion that all were deducible from a comparatively small number of postulates."[40] Bell wrote in a very engaging way, but he echoed false anecdotes, adding imagined details and exaggerations.[41]

Much earlier, the notion that Pythagoras *proved* the hypotenuse theorem spread thanks to the imaginative words of Galileo Galilei. In 1632, in his influential *Dialogue on the Two Chief World Systems*, Galileo wrote: "There is no doubt that a long time before Pythagoras had discovered the proof, for which he did the hecatomb, to be sure, that the square on the side opposite the right angle of the right-angled triangle was equal to the squares on the other two sides; and the certainty of the conclusion helps not a little in the discovery of the proof."[42]

But really, what are the earliest traces of the idea that Pythagoras was engaged in mathematics at all?

At around 480 CE, Joannes Stobaeus claimed that the musician Aristoxenus of Tarentum (ca. 320 BCE) once wrote that "Pythagoras seems to have honored, most of all, the study of numbers, and to have advanced it in withdrawing it from the use of merchants and tradesmen, likening all things to numbers."[43] Aristoxenus was a student of some Pythagoreans and of Aristotle, so his words seem to carry authority. Still, some historians doubt the accuracy of the quoted words because Aristoxenus could have just paraphrased Aristotle's claim that some so-called Pythagoreans advanced the study of numbers and also because such passages might denote a mystical numerology rather than any substantive mathematics.[44]

Next, about two centuries after Pythagoras died, a few writers, non-mathematicians, apparently claimed very briefly that he studied mathematics, among other things. For one, at around 50 BCE, Diodorus wrote that the Egyptians claimed that Pythagoras had learned from them "the art of geometry, arithmetic, and transmigration of souls."[45] It's a doubtful claim, considering that the ancient Egyptians did not believe in transmigration. Still, some of Diodorus's sources for his *Bibliotheca Historica* were copies of the writings of Hecataeus of Abdera (ca. 360–290 BCE), and hence some historians arbitrarily have ascribed the brief line in question to that early

Table 2. The Illusion of Knowledge

ca. 500 BCE	Pythagoras died.	If he wrote any works, none survived centuries later.
Four centuries later . . .		
45 BCE	Cicero	Pythagoras found something new in geometry and is said to have immolated an ox for the Muses; but I do not believe this.
15 BCE	Vitruvius	Pythagoras found that in a 3-4-5 triangle the square on side 5 sums the squares on the others, and allegedly he therefore sacrificed to the Muses.
Before 100 CE	Apollodorus	Pythagoras found the revered inscription, so he brought an ox-sacrifice.
ca. 100 CE	Plutarch	Pythagoras sacrificially burned an ox for his diagram: whether about hypotenuse power or the problem of the area of the parabola.
ca. 200 CE	Athenaeus	Pythagoras sacrificially burned a hecatomb upon finding that in the right triangle the hypotenuse equals in power its peripherals.
ca. 225 CE	Diogenes	Pythagoras sacrificed a hecatomb upon finding that in the right triangle the hypotenuse equals in power its peripherals.
ca. 300 CE	Porphyry	Pythagoras sacrificed an ox made of flour, having found that in the rectangle the hypotenuse equals in power its peripherals.
Thirteen centuries later . . .		
1632	Galileo Galilei's fictional dialogue	Pythagoras first knew that the square on the hypotenuse equals the squares on the triangle's other sides, and then he proved it and sacrificed a hecatomb.
1900s to the present	Thomas L. Heath, Eli Maor, Leonid Zhmud, and many other historians, writers, and teachers	Pythagoras *proved* the Pythagorean theorem.

The lack of specificity of Pythagoras's ancient achievements becomes increasingly disguised by later writers' apparent specificity and certainty over the centuries. Note: each claim is summarized; it is not a quotation.

source. However, the work of Diodorus, overall, suffers from so many flaws that one historian denounced him as one of "the two most accomplished liars of antiquity."[46]

Another apparently ancient but dubious claim that Pythagoras pursued geometry is a statement by Diogenes (ca. 225 CE), who reported that in *History of Alexander* (ca. 300 BCE) Anticlides claimed that the Egyptian king Moeris (Horemheb, ca. 1300 BCE) invented shadow-clocks and that afterward Pythagoras "advanced upon geometry"—words that are often mistranslated.[47] Here again, half of the sentence is dubious, as there is no evidence that King Moeris invented shadow-clocks.

Next, also around 250 BCE, Callimachus the poet reportedly said something about Pythagoras. Diodorus wrote:

> Callimachus, said about Pythagoras—because he found some of the problems in geometry, but others he first brought from Egypt to the Greeks—that:
>
> > "The Phrygian Euphorbus, who among men
> > found out about triangles and scalenes
> > and the circle in seven lengths,
> > taught to not eat living things:
> > but not everyone obeyed that."

However, the quoted poem by Callimachus does not specify Pythagoras, but merely "the Phrygian Euphorbus," so Diodorus construes this historically as an achievement of Pythagoras by saying that in a previous life the soul of Pythagoras had been in the body of Euphorbus, a legendary figure who supposedly died in the legendary Trojan War (ca. 1190 BCE). This is just imaginative fiction. It is also telling that the phrase "he found some of the problems in geometry, but others he first brought from Egypt to the Greeks" is not part of the poem quoted, but instead seems to be an interpolation by Diodorus.[48]

Claims have also arisen regarding the historian Herodotus. For example, in an otherwise excellent survey of ancient science, historian David Lindberg notes that "Herodotus (fifth century B.C.) reported that Pythagoras traveled to Egypt, where he was introduced by priests to the mysteries of Egyptian mathematics."[49] Not true: Herodotus made no such claim at all.

Finally, some writers claim that ancient coins connect Pythagoras to geometry. Silver coins from the ancient Greek city of Abdera (ca. 425 BCE) portray a bearded man framed in a square, surrounded by the name "PY-TAGORES." Specialists note that the name refers to the magistrate who issued the coins. Still, owing to the name's similarity to "Pythagoras," some writers guess that such images portray Pythagoras of Samos. For example, Christiane Joost-Gaugier speculates that the magistrate probably flattered himself by using the image of Pythagoras. She also speculates that the *square* frame in the *round* coin alludes to the *cube* of the Earth inscribed in the *sphere* of the universe, a geometrical formula.[50] But this speculation is unwarranted, given that various other ancient Greek coins from Abdera and elsewhere likewise framed heads in squares. By the way, many books show images that supposedly portray Pythagoras—ancient busts and coins—but I have found no evidence or credible explanations at all that warrant that any such image depicts Pythagoras of Samos.

What can we say about Pythagoras in the history of mathematics? Various ancient writers mentioned him, including Heraclitus of Ephesus, Plato, Herodotus, Heraclides, and Isocrates.[51] But those records do not claim that Pythagoras worked on mathematics. What do they say instead? We can briefly sum it up:

> Pythagoras was a popular religious leader who argued that the human soul is born repeatedly, even in animal bodies. He taught his followers to live in a disciplined way, including certain dietary restrictions, such as not eating beans or animals, at least of certain kinds.

To this we might add the impression of Aristoxenus: that Pythagoras seems to have honored and advocated numbers beyond their practical use.

In any case, Pythagoras became so popular that by 300 BCE onward, several forgers composed fake memoirs attributed to him and his followers. Several such texts still exist and have been identified as fake by historians and philologists.[52] Later alleged anecdotes are further obscured by the complication that there was not only one Pythagoras. For example, according to Pliny, there was a Pythagoras of Rhegium who was a famous sculptor, and there was another Pythagoras from Samos who was a famous painter.[53] And according to Diogenes Laertius, there were more: a Pythagoras who was a native of Croton and became a tyrant, another Pythagoras who was a trainer of wrestlers, another who was a native of Zacynthus, another who made statues in Rhodes and "is believed to have been the first to discover rhythm and proportion," another from Samos who also made statues, another who was a physician and writer, another who was an orator of no fame, and yet another who wrote a history of the Dorians; at least four of these men apparently lived at the same time as Pythagoras.[54]

In the end, what can we attribute to *the* Pythagoras with certainty in the history of mathematics? Nothing. As argued by historian Walter Burkert, "The apparently ancient reports of the importance of Pythagoras and his pupils in laying the foundations of mathematics crumble on touch, and what we can get hold of is not authentic testimony but the efforts of latecomers to paper over a crack, which they obviously found surprising."[55] Having researched the evidence above, I reach the same conclusion. Historian Otto Neugebauer briefly remarked that the stories of Pythagoras's discoveries "must be discarded as totally unhistorical" and that any connection between early number theory and Pythagoras is "purely legendary and of no historical value."[56] Over the centuries, the admirers of Pythagoras attributed to him multifarious political and miraculous feats, plus—why not?—scientific achievements. Writers and historians who still mathematicize Pythagoras rely mainly on speculations, old and new.

Looking back, we can surmise how the growth of myths about Pythagoras affected the history of mathematics. He was a charismatic religious leader. Many of his followers inflated his fame by adding tales of his alleged miracles and exploits in many fields, including evolving legends about his achievements in geometry and the sciences. They attributed to him ideas from Plato, Ptolemy, Jesus, and others: Pythagoras "healed the sick, raised the dead, stilled the waves of the sea with a word."[57] Eventually, people ceased to believe in the heroic miracle stories. Yet centuries later, the relatively new tales of his *intellectual* feats seemed ancient and therefore genuine. Math teachers groping for the history of their discipline adopted such tales as true, even if they appeared centuries or millennia after Pythagoras had died. Some historians embellish stories with rich conjectural details and inferences that "reconstruct" the past—guided by the intuition, invincibly acquired in childhood schooling, that Pythagoras was a great mathematician. It just sounds true.

Pythagoras still shines, in schoolbooks and science books, as a hero in the history of math. Even recent writers hardly resist the traditional portrayal. A beautifully illustrated book published by the Smithsonian claims, "This astonishing thinker and observer understood that the structure and relationships of the universe can be described with mathematical formulas. He made mathematics the language of Western science. No one has done more."[58] Eli

Maor, author of the popular and beautiful book *The Pythagorean Theorem: A 4,000-Year History,* admits that stories about Pythagoras "must be taken with a grain of salt"—but Maor still repeatedly glorifies Pythagoras: "Sometime around 570 BCE [*sic*], Pythagoras of Samos proved a theorem about right triangles that made his name immortal."[59] Likewise, in his book *God Created the Integers,* the bestselling physicist Stephen W. Hawking claimed that "I see no sufficient reason to question the tradition" that at least among the Greeks Pythagoras was the first to introduce and prove the hypotenuse theorem—but wait, actually, those are not Hawking's words: they are really the opinion of historian Thomas Heath, whose old commentaries on *The Elements* were reissued word for word, without attribution, as if Hawking had written them.[60]

Again, notice the contagious claim that Pythagoras *proved* the hypotenuse theorem. Yet there is not a single shred of evidence that he proved or ever even tried to prove that theorem or anything else in geometry. One historian fairly complains that "probably more sheer nonsense has been written about Pythagoras and his followers than about any other figure(s) in all the annals of mathematics."[61] But denials seldom have the charm and attractiveness of false reports. So how do we replace myths with history?

2

AN IRRATIONAL MURDER AT SEA

Fear prophets, Adso, and those prepared to die for the truth, for as a rule they make many others die with them, often before them, at times instead of them.
—Umberto Eco, *The Name of the Rose*

Long after Pythagoras died, some of his admirers were fascinated by numbers. This tradition eventually generated a captivating murder story.

If not Pythagoras himself, at least some of his admirers seemed to be interested in mathematics. Yet the earliest evidence is not complimentary. It suggests that some Pythagoreans focused on numbers not too thoughtfully. Plato criticized the Pythagoreans for analyzing numerically the harmonies of plucked strings, rather than analyzing relations among numbers themselves.[1] Aristotle repeatedly criticized "the so-called Pythagoreans" for believing that things, material things, are made of numbers.[2] And we too might think: "Things are made of *numbers*? Nonsense!" The ancient idea that things are made of numbers, so that numbers are *visible* things, not separate abstractions, seems weird. But it becomes plausible once we realize that many students nowadays do think that material objects are composed of mathematical things: three-dimensional geometrical figures, spherical subatomic particles.

Despite Aristotle's reports, we do not have documents in which any specific ancient Pythagorean claimed that objects are made of numbers. But there is evidence that at least some individuals did value numbers as an important property of things and valued our ability to think about numbers. For example, roughly a century after Pythagoras died, Philolaus reportedly said: "And indeed all things that are known have number. For it is not possible that anything whatsoever be understood or known without this."[3] This is the kind of ancient statement that mathematicians now appreciate, because it resembles our thinking: that to really know things it's important to think numerically. Philolaus also valued the role of simple ratios of whole numbers in the musical harmonies. By the way, contrary to later stories, there's no evidence that Pythagoras himself really discovered (or even knew about) such numerical ratios in the lengths of harmonic strings.[4]

Aristotle claimed that "the so-called Pythagoreans" had studied and developed the sciences and that having found numerical order in musical harmonies and other worldly things, they inferred "that the elements of numbers were the elements of all things, and that the whole heavens were harmony and number."[5] But Aristotle disagreed, and he criticized the Pythagoreans, for example, for inventing a heavenly body, a "counter-Earth," to match their numerical assumption about perfection, that there should be ten bodies in the heavens. There's also evidence that the Pythagoreans associated numbers not only with material things but with abstract notions. For example, Aristotle reported that they identified justice with the number four.

Centuries later, Iamblichus said: "that which is primary is the nature of numbers and ratios running through all things, according to which all these things are harmoniously arranged and suitably ordered."[6] Now, as it's difficult to abandon what we once learned, we might perhaps imagine that the reason why some of the admirers of Pythagoras valued numbers is likely because *he* initiated the religious study of numbers. However, in addition to the lack of evidence for this in the more ancient sources, there is evidence *against* this in the later sources. For example, Iamblichus claimed that Pythagoras derived his numerical worship from Orpheus, an ancient mythical figure, a musician and poet who charmed birds, beasts, and even rocks to dance and follow him. Iamblichus wrote that Pythagoras claimed to have learned from Agalophamus what Orpheus once declared, namely that "the eternal being of number is a most provident principle of the whole heaven, Earth, and of the intermediate nature; moreover, it is a source of permanence for divine men and gods and daemons."[7]

Anyhow, we don't know whether Pythagoras himself revered numbers in religion or life, but at least some of his later admirers did. And by "numbers," they did not mean what we mean: they did not include what we now call negative numbers, imaginary numbers, irrational numbers, infinitesimals, and more. Instead, it seems that they included only what we call the natural integers and ratios of them. This leads directly to a major legend about the Pythagoreans, namely that they were so fanatical in believing that *everything* is numerical that when Hippasus, a member of the cult, discovered that

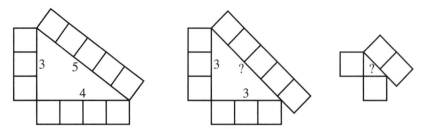

Figure 2.1. For a right triangle having side lengths of 3 and 4 units, the hypotenuse can be divided neatly into 5 units of the same size, so we call it "commensurable." But for right triangles having two sides of the same length, the hypotenuse is incommensurable, because no matter how small we make the squares, no integer quantity of squares fits both sides and the hypotenuse.

something could not be expressed as a ratio of numbers, the Pythagoreans killed him by drowning.

The story is often told as if Hippasus had been trying to express the hypotenuse of a triangle having two sides, each of 1 unit in length. Since the hypotenuse theorem states that $a^2 + b^2 = c^2$, then for a triangle with sides a and b, each of length 1, the length of the hypotenuse is $\sqrt{2}$. So it would seem that Hippasus was trying to find the numerical value of the square root of 2, that is, to express it as a ratio of two whole numbers. But he discovered that this was impossible and therefore that the Pythagorean religion *was wrong*. And he would not keep his mouth shut about this, so they murdered him.

Before analyzing this story, we should quote some of its manifestations. For example, a typical account was given by the mathematician Morris Kline in 1972, in a general history of mathematics across the centuries: "The discovery of incommensurable ratios is attributed to Hippasus of Metapontum (5th cent. B.C.). The Pythagoreans were supposed to have been at sea at the time and to have thrown Hippasus overboard for having produced an element in the universe which denied the Pythagorean doctrine that all phenomena in the universe can be reduced to whole numbers or their ratios."[8]

Readers learning basic history of math would look in a book such as Kline's and thus repeat what they read there. Some writers readily embellish the story by adding details or exaggerations. For example, here's a twisted account from 1997, in which Pythagoras himself murders his disciple: "Then one day, one of Pythagoras's disciples pointed out to him that the diagonal of a square whose side was one unit could not be expressed that way. . . . Since they were all on a boat at the time, Pythagoras threw his student overboard and swore everyone else in his class to secrecy."[9] But this tale is just the result of carelessly paraphrasing stories without checking the original sources. Here is a similar dramatic account:

> Hippasus must have been overjoyed by his discovery, but his master was not. Pythagoras had defined the universe in terms of rational numbers, and the existence of irrational numbers brought his ideal into question. . . . Pythagoras was unwilling to accept that he was wrong, but at the same time he was unable to destroy Hippasus's argument by the power of logic. To his eternal shame he sentenced Hippasus to death by drowning.
>
> The father of logic and the mathematical method had resorted to force rather than admit that he was wrong. Pythagoras's denial of irrational numbers is his most disgraceful act and perhaps the worst tragedy of Greek mathematics.[10]

Likewise, the bestselling book *Zero: The Biography of a Dangerous Idea* echoes the tale with cinematic details: "Hippasus of Metapontum stood on the deck, preparing to die. Around him stood the members of a cult, a secret brotherhood that he had betrayed. Hippasus had revealed a secret that was deadly to the Greek way of thinking, a secret that threatened to undermine the entire philosophy that the brotherhood had struggled to build. For revealing that secret, the great Pythagoras himself sentenced Hippasus to death by drowning. To protect their number-philosophy, the cult would kill."[11]

In other popular misstatements, Pythagoras himself makes the dreaded discovery. For example, in his Pulitzer Prize–winning book *Gödel, Escher, Bach*, Douglas Hofstadter duly notes that it was "Pythagoras, who first proved that the square root of 2 is irrational. . . . It was considered a truly sinister discovery at the time, for never before had anyone realized that there are numbers—such as the square root of 2—which are not ratios of integers. And thus the discovery was deeply disturbing to the Pythagoreans, who felt that it revealed an unsuspected and grotesque defect in the abstract world of numbers."[12] Likewise, in the bestselling book *Unknown Quantity: A Real and Imaginary History of Algebra*, John Derbyshire claims that "Pythagoras discovered to his alarm and distress" that the square root of 2 is irrational.[13] But such claims have no substance.

It is no wonder that just as such moonshine is published nowadays, when standards of scholarly rigor are well-known, other fictions were written many centuries ago, when standards and peer reviews were weaker or nonexistent. At around 460 CE, the pagan theologian Proclus apparently claimed that

Pythagoras "discovered the theory of irrationals." Decades ago, some historians used to take that claim seriously.[14] But historians later became increasingly skeptical of that claim, partly because it *only* appears more than *nine hundred years after* Pythagoras died. If it were true, why did none of Pythagoras's various biographers and commentators on the history of mathematics previously make any such claim? Instead, some historians speculate that even Proclus made no such claim, that his original words became corrupted.[15]

Even writers who carry out detailed research are not immune to the urge to speculate. For example, in 1945, the historian Kurt von Fritz wrote an article trying to show how Hippasus discovered irrationality.[16] Von Fritz argued that by drawing pentagrams inside pentagons, Hippasus could find that the side and diameter of a pentagon are incommensurable (i.e., no line segment, however small, can fit neatly into both the diameter and the side in integer multiples).

Von Fritz put together several bits of evidence to claim that this pentagon-pentagram procedure was the likely way in which Hippasus discovered that some lines cannot be represented by ratios of whole numbers. Von Fritz mentioned that a vase from the seventh century BCE shows a pentagram and that hence the Pythagoreans could have been well acquainted with that figure and could have studied its properties.[17] He also cited two writers who (centuries later) claimed that some Pythagoreans used the pentagram "as a token of recognition."[18] But such factoids are not at all evidence of what the Pythagoreans thought or did in the time of Hippasus. The bottom line is that von Fritz gave guesswork, not a historical finding. Still, some teachers have echoed von Fritz's argument as if it were history.[19] But really, how was

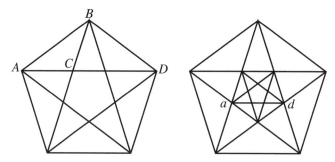

Figure 2.2. Can the lines *AB* and *AD* be divided neatly by a common length? No, by using the properties of triangles, geometers have shown that these two lines are incommensurable.

irrationality first discovered? We just don't know. Actually, we don't even know whether Hippasus discovered any such thing!

Von Fritz began his argument by claiming that historical tradition "is unanimous in attributing the discovery to a Pythagorean philosopher by the name of Hippasus of Metapontum."[20] But that was just not true at all. Von Fritz gave no example of anyone who attributed the discovery to Hippasus. Therefore, one historian objected that von Fritz's claim "seems to me to be devoid of all foundation. So far from being unanimous, the tradition is, I believe, non-existent. I know of no single ancient author attributing the discovery to Hippasus."[21]

What is the source of the now popular story about Hippasus, voiced by Kline, Von Fritz, and so many others? One earlier account appears in the work of John Burnet, in a book titled *Early Greek Philosophy*. Burnet mentioned that "our tradition says that Hippasos of Metapontium was drowned at sea for revealing this skeleton in the cupboard."[22] Burnet ascribed this story to Iamblichus. So what did Iamblichus say?

But first let's consider a less ancient source that mentions the discovery of irrationality. At around 340 CE, Pappus of Alexandria wrote a commentary on *The Elements,* in which he mentioned the disclosure of irrationality:

> Indeed the sect of Pythagoras was so affected by its reverence for these things that a saying became current in it, namely, that he who first disclosed the knowledge of surds or irrationals and spread it abroad among the common herd perished by drowning. Which is most probably a parable by which they sought to express their conviction that firstly, it is better to conceal (or veil) every surd, or irrational, or inconceivable in the universe, and secondly, that the soul which by error or heedlessness discovers or reveals anything of this nature which is in it or in the world, wanders (thereafter) hither and thither on the sea of non-identity (lacking all similarity of quality or accident), immersed in the stream of the coming-to-be and passing-away, where there is no standard of measurement. This was the consideration which Pythagoreans and the Athenian Stranger held to be an incentive to particular care and concern for these things.[23]

Several points are worth noting about this passage. First, Pappus believed that the story was likely a parable, not a historical fact. Second, the discoverer of irrationality allegedly died by drowning, but he was not murdered. And third, the story does not mention Hippasus at all.

Turn now to the prior writings of Iamblichus, at around 300 CE, where the other relatively ancient account of the discovery of irrationality appears:

> And Pythagoras is said to have taught first this very thing to those associating with him: that, free from all incontinence of will, they should guard in silence whatever discourse they heard. At any rate, he who first revealed the nature of commensurability and incommensurability to those unworthy to share in these doctrines was hated so violently, they say, that he was not only banished from their common association and way of life, but a tomb was even constructed for him. As one who had once been their companion, he had truly departed from life with human beings.
>
> Others say that even the Divine Power was outraged at those who divulged Pythagoras' doctrines. For that man perished at sea as an offender against the gods who revealed the construction of a figure having twenty angles: this involved inscribing the dodecahedron, one of the five figures called "solid," within a sphere. Some, however, maintained that the one who broke the news about the irrational and incommensurability suffered this fate.[24]

Figure 2.3. A dodecahedron inscribed in a sphere; two figures having twenty points of contact.

So, according to Iamblichus, there were two stories floating around about Pythagoreans who died at sea. One guy divulged the theory of incommensurables, and the other divulged a method for inscribing a twelve-faced figure (a dodecahedron) inside a sphere. Neither of them was murdered, except perhaps by their god, "the Divine Power." Or alternatively, in the other version of the story mentioned by Iamblichus, the guy who divulged the secret of incommensurables did not even die at sea, but the Pythagoreans erected a tomb for him, as if he were dead to them. Was Hippasus at least the guy for whom that tomb was made? Was he the divulger of the irrational?

Writers transform stories. For example, in his bestselling book *The Golden Ratio*, Mario Livio mistakenly claims: "According to one of Iamblichus' accounts, the Pythagoreans erected a tombstone to Hippasus, as if he were dead, because of the devastating discovery of incommensurability."[25] Historian Amir Alexander, in an otherwise excellent book, claims that Hippasus "proved" that the side and diagonal of a square are incommensurable.[26] Another writer portrayed him as a bold rebel: "Hippasos was a heretical Pythagorean . . . . He no longer felt bound by the anonymity in which other Py-thagoreans wrapped themselves. As often happens, the revelation concerned the Achilles tendon of Pythagorean thinking—the square root of two."[27]

But no. According to Iamblichus, Hippasus was the other guy: the one who revealed how to inscribe a dodecahedron in a sphere. Iamblichus wrote: "On the matter of Hippasus in particular: he was a Pythagorean, but because of having disclosed and given a diagram for the first time of the sphere from the twelve pentagons, he perished in the sea since he committed impiety. He acquired fame as having made the discovery, but all the discoveries were of that man, for so they refer to Pythagoras, and do not call him by his name."[28] Apparently the cult had the habit of attributing *all discoveries* to "that man," Pythagoras. Thus vanishes the legend that the discoverer of incommensurability was a Pythagorean called Hippasus, who was consequently murdered at sea. It is just a distortion of other stories. It originated by blending two mythical stories about death at sea. The story that a tomb was erected for Hippasus is also a misapprehension (see table 3).

Note also that the accounts of Iamblichus and Pappus do not specify whether the *discoverer* of incommensurability was a Pythagorean. But it would seem that at least the Pythagoreans knew about the topic and that they were awfully disturbed when one of them made that knowledge public. And their god too was outraged? So, even without the later embellishments, the story is engaging enough.

In any case, who was Hippasus? Well, aside from maybe divulging a way to inscribe a dodecahedron in a sphere, he seems to show up in a few more ancient sources. For example, Aristotle mentions a Hippasus who believed that the world was fundamentally made of fire. And Diogenes Laertius mentions that some authors believed that a Hippasus was Pythagoras's father or grandfather.[29]

Anyhow, we don't know who discovered incommensurability. But we do know that it was discovered prior to about 360 BCE, because at about that date, Plato wrote a dialogue that mentioned incommensurability: "Theodorus was writing out for us something about roots, such as the roots of three or five, showing that they are incommensurable by the unit: he selected other examples up to seventeen—there he stopped."[30] Since Theodorus discussed incommensurability, some writers speculate that he was a Pythagorean or that he learned it from a Pythagorean.[31] And how did the ancients first find that some lengths are incommensurable with others? We don't know, but at least we have an allusion to an early proof. Aristotle made

Table 3. The Blending of Stories

ca. 400? BCE	Hippasus lived.	If he wrote any works, none survived centuries later.
Six centuries later . . .		
ca. 180? CE	Lucian	The Pythagoreans used a pentagram as a token of recognition.
ca. 300	Iamblichus	Hippasus was a Pythagorean who first revealed how to inscribe a figure of twelve pentagons into a sphere, and he died at sea for committing impiety.
ca. 300	Iamblichus	Someone who first revealed incommensurability to the unworthy was hated so violently, they say, that he was banished and a tomb was constructed for him. Some others say instead that this person died at sea as an offender against the gods.
340	Pappus	A saying or parable claims that he who first disclosed the knowledge of irrationals and spread it among the common herd died by drowning.
Fifteen centuries later . . .		
1892	John Burnet	Tradition says that Hippasos was drowned at sea for revealing irrationality.
1945	Kurt von Fritz	Hippasus discovered irrationality, probably by analyzing the lines of a pentagram.
1980	James R. Choike	Kurt von Fritz showed that Hippasus discovered irrationality in the lines of a pentagram.
1972	Morris Kline	The discovery of incommensurable ratios is attributed to Hippasus, and the Pythagoreans supposedly threw him overboard.
1997	William Everdell	One of Pythagoras's disciples pointed out that the diagonal of a square is incommensurable, so then "Pythagoras threw his student overboard and swore everyone else in his class to secrecy."
2005	Stephen Hawking	Whoever he was, the man who discovered irrationality "was the first martyr for mathematics!"
2006	John Derbyshire	"Pythagoras discovered to his alarm and distress" that the square root of 2 is irrational.
2006	Mario Livio	"The Pythagoreans erected a tombstone to Hippasus, as if he were dead, because of the devastating discovery of incommensurability."
2010	Amir Alexander	"Hippasus of Metapontum proved that the side of a square is incommensurable with its diagonal."

A story about Hippasus becomes conflated with a story about a Pythagorean who revealed irrationality, and writers add imagined details and conjectures. Note: unless quotation marks are used, each claim is summarized.

a passing mention of how one can find incommensurability: "For all who argue by impossibility infer by syllogism a false conclusion, and prove the original conclusion hypothetically when something impossible follows from a contradictory assumption, as, for example, that the diagonal [of a square] is incommensurable [with the side] because odd numbers are equal to even [numbers] if it is assumed to be commensurate."[32] Aristotle's words resemble a procedure given later in *The Elements.* It's a wonderful argument that we may clarify as follows.

Consider a square having sides of length 1 each. By the hypotenuse theorem, its diagonal is then the square root of 2. Is there any segment of length, however small, that can be used to neatly divide both the diagonal and a side of the square? If so, then the length of the diagonal can be expressed as a ratio, stating how many units fit in the diagonal and how many fit in the side. So, the argument proceeds by assuming that the square root of 2 really is a ratio and consequently deducing an absurd contradiction, which then falsifies the assumption.

So, suppose that $\sqrt{2}$ can be expressed a ratio of integers, that is, that

$\sqrt{2}=c/d.$

If c and d have any common factors, then we should reduce them to their simplest form. For example, if $c = 14$ and $d = 10$, then, since both 14 and 10 are neatly divisible by 2, we can simplify the fraction to 7/5, and now the ratio is in its simplest form. Thus the ratio c/d can be reduced and expressed in its simplest form, which we may call a/b, where both a and b are again whole numbers and b is not zero. So we write: $\sqrt{2}=a/b$. It follows that

$a^2 = 2b^2$

Therefore, a^2 is an even number since it is two times something. So, a itself is also even, because all even squares have even square roots. And, since a is even, then b must be *odd*, because we started by simplifying a/b, such that they were not both even (otherwise they could still be divisible by 2).

Now, since a is even, a is twice some other whole number, say, $a = 2k$. If we substitute $a = 2k$ into the original equation $\sqrt{2}=a/b$ we get $2=(2k)^2/b^2$, which means that

$b^2 = 2k^2$

This means that b^2 is even, from which it follows again that b itself is an *even* number. But it can't be! That's a contradiction, because we found above that it had to be odd. It is impossible that b is both even and odd. But since we reached that conclusion by assuming that $\sqrt{2}=a/b$, then we conclude that this assumption is false. We conclude that $\sqrt{2}$ *cannot* be a ratio of integers.

So this is the kind of argument that Aristotle and others gave to prove that the diagonal of a square is incommensurable with its side.

So what? Why does it matter that there doesn't exist one unit of length that divides both the side and the diagonal of a square? It doesn't seem to have been of much importance to Aristotle. But according to Pappus and Iamblichus, it did matter to the Pythagoreans. Since numbers were just integers and their ratios, it meant that there was no number in existence that corresponded to the diagonal of a square. It would seem that $\sqrt{2}$ was not a number. It meant that the problem of extracting the square root had no solution, no numerical answer. It meant that the idea that Aristotle attributed to the Pythagoreans, that numbers are the principles of all things, was wrong.

Thus, indeed, it may seem that some individuals might have felt annoyed and desired to keep incommensurability secret. A number that was not a ratio was not a number; it would be inconceivable, a crazy paradox. And now, if someone is insane, we say that they are irrational.

Nowadays, we do not suffer the diagonal of a square. We simply say that it can be represented by a particular kind of number, an irrational number. But in Greek antiquity irrational magnitudes were not numbers. And the reportedly Pythagorean theory that all is number was ridiculed by Aristotle. Afterward, Euclid synthesized most geometers' contributions into the *Elements*. The author or authors of the *Elements* took pains to formulate everything geometrically, even where numerical or algebraic arguments would be simpler. The *Elements* ends with the construction of the five regular solids: those having four sides, six, eight, twelve, and twenty. Decades earlier, Plato had argued that reality, the invisible but eternal world of forms, was composed of these five regular solids.

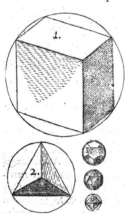

Figure 2.4. The five regular solids, as drawn by Johannes Kepler in his *Harmonices Mundi* of 1619.

So there was, since antiquity, a tension between those who thought that the most fundamental mathematical entities are numbers and those who preferred geometrical figures. Through the centuries, these viewpoints recurred, as some mathematicians sought to explain numerical notions geometrically, while others tried to arithmetize geometry.

Thus, if you happen to be the kind of mathematician who prefers numbers over shapes, then the so-called Pythagoreans will seem heroic. Again, the distant past resonates to the extent that it echoes our likes.

Alternatively, we can well disagree with the Pythagoreans, say, for their secrecy and their readiness to ascribe to Pythagoras achievements that were not his (as was their custom, according to Porphyry and Iamblichus).[33] And we can also reject their claim that it is unlawful to reveal true knowledge to anyone in an easily intelligible way.[34]

Returning to the legends about Pythagoras's triangle and Hippasus at sea, we can surmise why these stories propagated so widely. Both legends have something in common. In both, some zealots valued mathematical knowledge as being so important and sacred that they were ready to kill for it. Such legends sound true because they match some of the notions and biases that we acquire from popular culture. They echo the popular stereotype that secretive pagan cults carry out sacrifices, conspiracies, and extreme retributions. The Pythagorean legends also seem to confirm the idea that genuine mathematics began in the Hellenistic civilization of the Mediterranean, in and around ancient Greece. And above all, these legends convey the idea that mathematical knowledge is valuable. *It is,* but legends are unnecessary to convey that truth.

Nevertheless, the myth of Hippasus is *valuable* because it enables us to meditate on a contradiction. Even in its most extreme version, that Pythagoras murdered his disciple, the story is engaging because it moves us to reflect on the idea of somebody who abuses his power, a man distinguished for his rationality who unfairly acts in an irrational way. Likewise, the group of Pythagoreans allegedly stood by: they allowed the creative heretic to be punished. Rationality was carried to the extreme. This myth challenges us, teachers and mathematicians, to not behave like the cult of Pythagoras.

At least stories sometimes regress toward earlier forms. For example, the bestselling physicist Stephen Hawking does not claim that the discoverer of irrationality was Hippasus. But still, Hawking claims that the Pythagoreans plotted to drown the discoverer at sea and that "whoever he was, this man was the first martyr for mathematics!"[35]

We should fight the habit of perpetuating familiar myths about who was the first to do this or that. Instead, we can learn from these Pythagorean tales how easily hearsay and fictions contaminate education by masquerading as history. It is not only bad that we learn myths as history. Moreover, the myths and legends have been distorted. So we might agree with the Pythagoras of

Iamblichus, as allegedly "he condemned both prose writers and poets for the errors in their versions of myths."[36] But the problem is that, unlike history, legends often *improve* owing to distortions and errors.

Good stories evolve. In 1632 Galileo published his *Dialogue* on the Earth's motion, despite fair warning that this "Pythagorean" theory was unwelcome by the Catholic Church. He presented it as a hypothesis in a fictional dialogue, to not offend. And he told this story about the Pythagoreans: "But the mysteries for which Pythagoras and his sect held in such veneration the science of numbers, it being nonsense what is spread by the mouths and writings of the vulgar, I do not believe in any way: indeed because I know that they did not expose wonderful things to the ridicule and contempt of the populace, to damage, as sacrilege to publish the most recondite properties of numbers, and about the incommensurable quantity & irrational which they investigated, and they preached that the one who had revealed this was tormented in the other world."[37] Apparently it was not enough that Hippasus and the one who revealed irrationality died at sea. That finite punishment did not suffice, so in Galileo's version it seems that anyone who committed such crimes would be tortured in hell.

3

UGLY OLD SOCRATES ON ETERNAL TRUTH

There is no good evidence that Pythagoras linked mathematics and religion, but apparently someone else did. Socrates lived in Athens in the fifth century BCE. According to ancient accounts, he was very ugly, with bulging eyes and a flat, upturned nose with wide-open nostrils. Allegedly he became a soldier and fought bravely in some battles. In old age he was bald, poor, constantly barefoot, and spent much time discussing philosophy in the city streets, questioning men who presumed to be wise, criticizing their ideas, annoying them by showing the contrary. Although we do not have any works by Socrates, some of his followers wrote about his views. One of them recounted how someone allegedly once described ugly old Socrates: "You exist, I do not say live, in a style such as no slave serving under a master would put up with. Your food and drinks are of the cheapest sort, and as to clothes, you cling to one wretched cloak which serves you for summer and winter alike; and so you go around all throughout the year, without shoes on your feet or a shirt on your back. Then again, you are not taking or making money."[1] Reportedly, Socrates replied that by living modestly, he gained independence and that he did not prostitute wisdom by selling it, that he instead gave it freely in friendship.

Another one of Socrates's followers, Plato, wrote several dialogues that also apparently convey his mentor's arguments. According to an ancient account, young Plato's wrestling coach had given him the name "Platon," meaning "broad," because he was a big guy. Some older accounts say instead that his name came from the breadth of his eloquence or because he had a broad forehead. In any case, Plato became more interested in philosophy than in headlocks, and his dialogues about Socrates discussed ideas about mathematics that became extremely influential.

In one of Plato's dialogues, Socrates describes himself as an honest man who doesn't care about what many people care about: wealth, family interests, military titles, and so forth.[2] In another dialogue, old Socrates speaks with Meno, a handsome, young, and educated man from Thessaly, Greece.[3] Meno asks him how virtue arises, whether it can be taught or acquired, but Socrates replies, "I know literally nothing about virtue," and he gradually convinces Meno that since people do not know what virtue is, there are no teachers of virtue. Along the way, Socrates speaks about geometry. The conversation between Meno and Socrates shows elements of what became famously known as Plato's philosophy of mathematics, an outlook later attributed to Pythagoras.

While trying to define virtue, Meno becomes exasperated with Socrates for "always doubting yourself and making others doubt; and now you are casting your spells over me, and I am just getting bewitched and enchanted, and am at my wits' end." He complains that Socrates has shocked him, like an electric ray fish, saying that in a city other than Athens, Socrates "would be cast into prison as a magician." Socrates admits that he too does not know much about virtue, but that he has heard of wise priests and priestesses who speak of divine things, the human soul: "The soul, then, as being immortal, and having been born again many times, and having seen all things that exist, whether in this world or in the world below, has knowledge of them all; and it is no wonder that she should be able to call to remembrance all that she ever knew about virtue, and about everything; for as all nature is akin, and the soul has learned all things," and therefore all knowledge is really recollection.

To explain what he means, Socrates asks Meno to bring one of his many servants, a slave, to answer some questions, to show Meno whether the answers come from learning or from remembering. A slave boy joins them, one uneducated in geometry. Socrates shows him a square:

The slave boy recognizes that this square has an area of four units. Socrates asks by how much the area would be increased if the length of the side of the square were doubled. The boy guesses that the larger square would have an area twice as large: eight units.

But Socrates then asks: "But does not this line become doubled if we add another such line here?"

Yes, and by describing a large square on the basis of that doubled length, Socrates helps the boy realize that by comparison to the smaller square, the bigger square is not double the area: it's more than eight square feet:

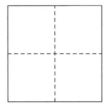

By doubling the size of the sides, the boy realizes, the area of the square becomes *four* times greater. The question remains: given the original square, how can we construct a square area just *twice* as large?

Through questions, Socrates shows the boy that a square twice as big would have sides larger than two feet and smaller than four feet—and not side lengths of three feet, either, because then its area would be nine square feet instead of eight. To illustrate the problem, Socrates again draws the square of four units and adds two identical squares to it:

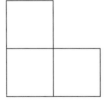

By filling up the vacant corner, the boy recognizes that the resulting area is four times larger than the initial square. Next, Socrates cuts the original square by using a diagonal line:

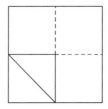

The boy realizes that the line divides the square into two equal parts. Then Socrates traces more lines on the diagram:

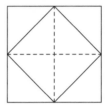

The boy analyzes the internal figure:

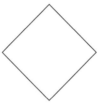

It is built from the diagonal of the original smaller square. Socrates says: "And that is the line which the learned call the diagonal. And if this is the proper name, then you, Meno's slave, are prepared to affirm that the double space is the square of the diagonal?" Yes, the boy agrees, this square doubles the area of the first.

Socrates then tells Meno that the boy had not been taught these facts about geometry, but that the knowledge was inborn and had been recovered by answering questions. Socrates says that geometrical truths existed in the soul of the boy long before he was born.

As portrayed by Plato, mathematics is eternal, such that its knowledge is not made but remembered by those who pursue it. Socrates tries not to tell the boy geometric relations, to thus challenge him to recognize the truth for himself. Socrates argues that students, better than thinking they know when really they don't, can benefit from confusion: they are better off knowing their ignorance.

Plato wrote about math in other dialogues. His *Republic,* composed around 375 BCE, argues that geometry and arithmetic are essential in education. This dialogue compares humans to prisoners in a cave, prisoners who have not yet seen reality, distracted by illusions, and Plato argues that mathematics leads our minds to truth.[4]

Socrates portrays humans as prisoners in an underground cave since childhood, their legs and necks chained to prevent them from turning around to view the source of light at the mouth of the cave. The prisoners can only see images on a wall, images that they think are real but that are really just shadows projected from behind, shadows of things between them and the source of light. Socrates says that if a prisoner were freed and dragged out to see sunlight, it would be painful and irritating, and it would take the prisoner time to manage to see real things. Among such realities are the elements of mathematics.

In order to "rise out of the sea of change, and grasp true Being," Socrates says, military men and philosophers alike should learn arithmetic, because the study of numbers leads to truth:

> This kind of knowledge may be prescribed by legislation; and we must try to persuade those designated as leaders of our State to go and learn arithmetic, not as amateurs, but they must carry on the study until they see the nature of numbers with the mind only; nor again, like merchants or retail-traders, with a view to buying or selling, but for the sake of their military use, and of the soul herself; and because this will be the easiest way for her to pass from Becoming to truth and Being. . . . I mean, as I was saying, that arithmetic has a very great and elevating effect, compelling the soul to reason about abstract number, and rebelling against the introduction of visible or tangible objects into the argument. You know how steadily the masters of the art repel and ridicule anyone who tries to divide absolute unity when he is calculating, and if you divide, they multiply, taking care that one shall continue one and not become lost in fractions.

Socrates says that we should study the "wonderful numbers" (the integers), which all have unity, each unit being "equal, invariable, indivisible," numbers that can be grasped by thinking. He calls this kind of knowledge *necessary,* based on the use of "pure intelligence to attain pure truth."

Glaucon replies that the citizens who defend the state should learn both arithmetic and geometry, especially "that part of geometry which relates to war; for in pitching a camp, or taking a position, or closing or extending the lines of an army, or any other military maneuver, whether in actual battle or on a march, it will make all the difference whether a general is or is not a geometrician." Socrates agrees but explains that a little geometry suffices for military uses, but that more advanced geometry enables students to envision the idea of good, which "compels the soul to gaze toward that place where is the full perfection of Being."

Socrates says that geometry is more important if it focuses on Being, and he complains that this proper conception of geometry contradicts the ordinary words used by geometers: "They regard practice only, and are always speaking in a narrow and ridiculous manner: of squaring and extending and applying and the like—they confuse the necessities of geometry with those of daily life, whereas knowledge is the real object of the whole science." Glaucon agrees, and Socrates adds that "the knowledge at which geometry aims is knowledge of the eternal, and not of anything perishing and transient." Geometry will lead students and citizens to truth, so Socrates requires that "nothing should be more sternly laid down than that the inhabitants of your fair city should by all means learn geometry."

Socrates and Glaucon next discuss whether astronomy should be the next subject of study for citizens. Glaucon endorses this proposal because knowing the seasons, months, and years would be useful to everyone, from farmers to sailors. But Socrates comments:

> I am amused, at your fear of the world, which makes you guard against the appearance of insisting upon useless studies; and I quite admit the difficulty of believing that in every man there is an eye of the soul which, when by other pursuits is lost and dimmed, is by these purified and re-illuminated; and is far more precious than ten thousand bodily eyes, for by it alone is truth seen. Now there are two kinds of persons: one kind who will agree with you and will take your words as a revelation; another kind to whom they will be utterly meaningless, and who will dismiss them as idle tales, for they see no sort of profit in them.

Socrates therefore argues that calculation and geometry and other elements of education should be presented in childhood. But he cautions that they should not be forced onto students as a kind of slavery, because "knowledge acquired under compulsion gains no hold on the mind." Instead, he suggests that math should be presented as a kind of amusement.

The speakers in the *Republic* stress the importance of math, that it is an important aspect of being human. They argue that soldiers, philosophers, and artists must all learn arithmetic. Socrates asks: "Can we deny that a warrior should have a knowledge of arithmetic?" Glaucon replies: "Certainly he should, if he is to have the smallest understanding of military tactics, or indeed, I should rather say, if he is to be a man at all." In short, they say:

1. All arts and sciences involve numbers and calculation, which have universal application.
2. Arithmetic is a truly necessary knowledge.
3. The "true use" of arithmetic is to draw the soul toward Being.
4. Geometry aims at knowledge of the eternal.
5. More than anything, geometry lifts the soul toward truth.

For centuries geometrical knowledge was valued as a divine treasure. Consequently, since some Italian followers of Pythagoreans became known for having religious views on mathematics, some writers speculated that Plato was a follower of the Pythagoreans. One cause for this conjecture is a passage in which Aristotle noted that Plato's philosophy came *after* some Italian schools of thought.[5] Needlessly, writers speculated that Plato "followed Pythagoras."

Did Socrates value mathematics as an end in itself? Given that Plato's *Republic* portrays Socrates advocating the importance and priority of mathematics in guiding the soul toward eternal truth, we might well wonder what length of that journey he recommended one take in terms of sophisticated mathematical steps. According to another of Socrates's followers, Xenophon, Socrates argued that although basic geometry is very valuable, the extended pursuit of abstruse geometry seemed hard to justify:

> Everyone (Socrates would say) should be taught geometry so far, at any rate, as to be able, if necessary, to seize or yield a piece of land, or to divide it or assign a portion for cultivation, and in every case by geometric rule. That amount of geometry was so simple and easy to learn, that it only needed ordinary application of the mind to the method of mensuration, and the student could ascertain the size of the piece of land, and with the satisfaction of knowing its measurement, depart in peace. But he was unable to approve of the pursuit of geometry up to the point at which it became a study of unintelligible diagrams. What the use of these might be, he failed to see, he said; and yet he was not unversed in these recondite matters himself.

These things, he would say, were enough to wear out a man's life, and to hinder him from more useful studies.[6]

So perhaps Socrates endorsed only a basic mathematics, as a means to live well and to train the mind to find truths. For him, apparently, geometry was not an end in itself, but a prerequisite for training the mind to develop abstract thinking, beyond what is visible. Nevertheless, Plato's dialogues later generated a widespread impression that mathematics should be pursued *independently* of its practical applications.

In 399 BCE, a young poet accused old Socrates of the capital crime of irreverence toward the gods of Athens. Allegedly Socrates's critical comments about some of the presumed traits of the gods had corrupted the beliefs of young men, so the city officials put Socrates on trial. In one dialogue, Plato refers to himself as one of the young men whom Socrates knew and who was present at the trial. Socrates had opportunities to contest the accusations and also to leave the city, but he did not. After the trial, the jury found him guilty. Socrates refused to beg or cry for his life; he instead discussed what kinds of punishment or fine might be suitable for his alleged crime, aside from death. But the magistrates remained unconvinced: they required a death sentence. Plato and other friends offered to pay a fine, but their offer was not accepted. Socrates also made a prediction about those who condemned him:

> I am about to die, and that is the hour in which men are gifted with prophetic power. And my prophecy to you who are my murderers, is that immediately after my death punishment far heavier than you have inflicted on me will surely await you. You have killed me because you wanted to escape the accuser, and not to give account of your lives. But that will not be as you expect: far otherwise. For I say there will be more accusers of you than there are now; accusers whom hitherto I have restrained; and as they are younger they will be more severe with you, and you will be more offended at them. For if you think that by killing men you can avoid the accuser censuring your lives, you are mistaken.[7]

Reportedly, ugly old Socrates faced death without fear and spoke about the immortality of the soul. He argued and hoped that death is good, rather than evil. He asked his friends to someday punish his accusers and judges, not because they had harmed him but because they had not meant to do him any good; because they cared more about wealth than about virtue; and because they pretended to be something when really they were nothing.[8] Socrates was imprisoned and chained. Friends encouraged him to escape from the city, offering help, but he stayed; he argued that he had agreed to live by the laws of Athens.[9] Later, Socrates bathed at the prison so that the women of his family would not have to wash his filthy corpse. The jailer said goodbye, calling Socrates "the noblest, the gentlest, and the best man of all that ever came here." Then the jailer cried. An executioner prepared and served a beverage poisoned with hemlock. Old Socrates cheerfully took the cup, prayed to the gods, and drank the poison. His friends wept, but he told them to be quiet, to let him die in silence. The poison stiffened and numbed his legs, affected his torso, and killed him.[10]

Years later, Plato established a school in Athens, called the Academy (ca. 387 BCE), reportedly in an open grove of olive trees. The school was soon enclosed by walls to protect it. A late tradition claims that at the entrance of the Academy, these words were inscribed: "Let no one inapt to geometry enter here."[11] Apparently the Academy thrived for hundreds of years, but it ceased operating around 83 BCE. It was reestablished in 410 CE, but finally in 539 CE the Roman emperor Justinian of Byzantium took control of it and essentially shut it down. Justinian was a Christian, and apparently he wanted to stop the Hellenic pagan influence of the Academy. Yet Plato's Academy was such a successful and famous school that we should look further at Plato's outlook on mathematics and education.

Plato's *Republic* suggests that physical perceptions, the world of experience, are illusory. At most, this world is a shadow of the eternal reality, the world of forms. Plato's dialogues became so influential that readers and philosophers extracted some recurring themes and embodied them into a viewpoint that became known as "Platonism," though it is unclear whether Plato himself actually held such views or to what extent. According to Platonism, mathematical knowledge is eternal, universal, unchanging, and fundamental to education and reasoning. It would then seem that mathematics is immune to the ravages of history: humans might not know all mathematics at one time, but nothing can be changed in mathematics; that is, politics, revolutionary wars, or misery will never change the truth of any mathematical proposition. Moreover, according to Platonism, physical objects and processes cannot be used to prove propositions in geometry. Geometry is *independent* of physics. For example, given a triangle inscribed in a rectangle, as illustrated, what is the ratio of this triangle to the rectangle?

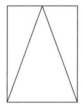

Clearly, the ratio is 1:2, the triangle occupies half the area of the rectangle. Now, given instead a cylinder and a cone inscribed in the cylinder, what is the ratio of the volume of the cone to the volume of the cylinder?

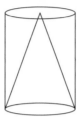

The answer is not self-evident: we might look in a geometry book or consult a proof to find the answer. For example, there is a proof in the *Elements*, book 12, Proposition 10.

But is there a simpler way of finding out this answer? One simple way is to construct the cylinder out of paper, craft the cone out of paper too, fill up the cone with water, pour that water into the cylinder, and then measure how much it fills up. If this is done very carefully, the cylinder fills up to what very much seems to be one-third.

However, the arguments voiced in Plato's dialogues suggest that we should reject this watery demonstration as totally unacceptable. We used material things, all of them imperfect. The rolled paper, for example, looks like a cylinder, but it's not really a cylinder, because its sides are not perfectly smooth. How can one gain knowledge of the universal by using particular things, imperfect things that bend, break, and rot? Consider a comment by the historian Plutarch, composed around 100 CE, about Plato's disdain for physical procedures in geometry:

> Eudoxus and Archytas had been the first originators of this far-famed and highly-prized art of mechanics, which they employed as an elegant illustration of geometrical truths, and as means of showing experimentally, to the satisfaction of the senses, conclusions too intricate for proof by words and diagrams. . . . But what about

Plato's indignation at it, and his invectives against it as the mere corruption and annihilation of the one good of geometry, which was thus shamefully turning its back on the un-embodied objects of pure intelligence to recur to sensation, and to ask help from matter (not to be obtained without base supervisions and depravation); so it was that mechanics became separated from geometry, and, repudiated and neglected by philosophers, took its place as a military art.[12]

Likewise, Plato's most famous student, Aristotle, claimed that it is wrong to try to use knowledge from physics to prove things in mathematics. Aristotle argued: "It is not for one science to prove something belonging to a different science, except when the things are so related that one is subordinate to the other, that is to say, e.g. theorems in optics are to geometry, and theorems in harmonics are to arithmetic."[13] Math could be used to prove statements in the sciences, but not vice versa. So here are some ancient roots of the common notion, still prevalent today, that mathematical proofs must be pure, devoid of physical constructions.

Plato's outlook on mathematics was immensely influential, perhaps the most influential in history. Without necessarily labeling anyone "a Platonist," we can readily find prominent individuals who voiced views reminiscent of some key aspects of Plato's dialogues. At around 450 CE, for example, Proclus emphasized the idea that geometry lifts the soul beyond material concerns, though he ascribed such views to the cult of Pythagoras: "I emulate the Pythagoreans who even had a conventional phrase to express what I mean: 'a figure and a platform, not a figure and a sixpence,' by which they implied that the geometry which is deserving of study is that which, at each new theorem, sets up a platform to ascend by, and lifts the soul on high instead of allowing it to go down among the sensible objects and so become subservient to the common needs of this mortal life."[14] Likewise, in 1570 the Christian magician John Dee, an advisor of Queen Elizabeth I, advocated Plato's views of mathematics in a long preface to an English translation of the *Elements*. He argued that

> it must be confessed (said Plato) that Geometry is learned for the knowing of that which is ever, and not of that which in time both is bred and is brought to an end, etc. Geometry is the knowledge of that which is everlasting. It will lift up therefore (O gentle Sir) our mind to the Verity: and by all means, it will prepare the Thought to the

Philosophical love of wisdom: that we may turn or convert toward heavenly things (both mind and thought) which now, otherwise than become us, we cast down on base or inferior things, etc.[15]

John Dee viewed mathematics as entwined with practical matters, such as architecture and navigation, but also with mystical beliefs, such as astrology. He believed that mathematics was central to the advancement of knowledge. Dee requested that London should issue a commandment stating that inhabitants of the city should not hate geometry.[16]

The German mathematical astronomer Johannes Kepler also connected mathematics and religion. In 1619, in his book *The Harmony of the World*, Kepler argued that geometry was not created by humans, as he conveyed its divine importance: "Geometry existed before the creation; is co-eternal with the mind of God; is God himself. . . . Where there is matter there is geometry. . . . Geometry provided God with a model for the Creation and was implanted into man, together with God's own likeness—and not merely conveyed to his mind through the eyes. . . . It is absolutely necessary that the work of such a Creator be of the greatest beauty."[17] Mathematicians valued numbers and arithmetic with similar certainty. For example, in the 1890s, Charles Hermite argued that "the (whole) numbers seem to me to be constituted as a world of realities which exist outside of us with the same character of absolute necessity as the realities of Nature, of which knowledge is given by our senses."[18] Consider another example: in 1964, the mathematician and logician Kurt Gödel defended the reality of the mathematical objects of set theory: "Despite their remoteness from sense experience, we do have something like a perception also of the objects of set theory, as is seen from the fact that the axioms force themselves on us as being true. I don't see any reason why we should have less confidence in this kind of perception, i.e., mathematical intuition, than in sense perception. . . . They, too [the objects of set theory], may represent an aspect of objective reality."[19] Gödel actually described himself as a Platonist, and he construed "Platonism or 'Realism'" as the view that "mathematical objects and facts (or at least something in them) exist independently of our mental acts and decisions" and that "the objects and theorems of mathematics are as objective and independent of our free choice and our creative acts as in the physical world."[20]

Next, consider claims made by the Russian mathematician I. R. Shafarevitch. In 1973, receiving a prize from the Academy of Science at Göttingen, West Germany, Shafarevitch explicitly connected mathematics to religion:

A superficial glance at mathematics may give an impression that it is a result of separate individual efforts of many scientists scattered about in continents and in ages. However, the inner logic of its development seems much more the work of a single intellect, developing its thought systematically and consistently using the variety of human individualities only as a means. It resembles an orchestra performing a symphony composed by someone. A theme passes from one instrument to another, and when one of the participants is bound to drop his part, it is taken up by another and performed with irreproachable precision. . . . One is struck by the idea that such a wonderfully puzzling and mysterious activity of mankind, an activity that has continued for thousands of years, cannot be a mere chance—it must have some goal. Having recognized this we inevitably are faced by the question: *What is that goal?* . . . I want to express a hope that . . . mathematics may serve now as a model for the solution of the main problem of our epoch: to reveal a supreme religious goal and to fathom the meaning of the spiritual activity of mankind.[21]

Furthermore, rather than quote only famous mathematicians, it is interesting to quote also the views of a math teacher, someone well qualified but not internationally renowned. An interesting example appears in the book *The Mathematical Experience,* by Philip Davis and Reuben Hersh. In April 1978, someone interviewed the chair of the Mathematics Department of a fine private school in New England. He had a master's degree in mathematics from an Ivy League university, and he taught mathematics, physics, and general science at his prep school. He said that the history and philosophy of mathematics did not arise in his classes. Asked whether mathematics is discovered or invented, he answered: "There's not much difference between the two. Why waste time trying to figure it out? The key thing that is important is that doing math is fun. That's what I try to put across to the kids."

But the interviewer insisted on the question about discovery or invention. The teacher then replied: "Well, I think it's discovered."

Asked if he had ever thought about the consistency of mathematics, he replied:

"I've heard about the Russell paradox and all that, but I really don't understand it. I think math is like a sandcastle. It's beautiful, but it's made of sand."

"If it's made of sand, how do you justify its study to your students?"

"I tell them that figures don't lie. You know they don't. No one has come up with any counterexamples to show that they do. But the whole question is irrelevant to me."

In answer to the question as to whether there is a difference between pure and applied mathematics, he answered, "Pure math is a game. It's fun to play. We play it for its own sake. It's more fun than applying it. Most of the math that I teach is never used by anyone. Ever. There's no math in fine arts. There's no math in English. There's no math to speak of in banking. But I like pure math. The world of math is nice and clean. Its beautiful clarity is striking. There are no ambiguities."[22]

This teacher's views on mathematics have some resemblance to Platonist views. As in the dialogues of Socrates, math seems to exist independently of humans, it is not an invention, and both Socrates and this teacher argue that mathematics should be taught in a way that is enjoyable. Yet there is something willfully thoughtless in this teacher's views on math. The interviewer asked him: "Does the number pi exist apart from people? Would the little green man from Galaxy X-9 know about pi?" He replied: "As one gets older, one is less and less inclined to trouble oneself about this kind of question." Knowing that he was teaching computer programming, the interviewer asked: "What's the purpose of computing?" He replied: "No one in high school asks 'why.' It's there. It's fun."

The comments of this teacher are remarkable. He certainly voiced some thoughts similar to the Platonist philosophy of math. Yet his intentional thoughtlessness is quite a departure from Socrates.

4

THE DEATH OF ARCHIMEDES

The idea that geometry is timeless led people to think that there can be no change in mathematics. There can be discovery, they thought, but not invention. It also encouraged the idea that change and moving things are foreign to pure mathematics. Euclid seemed to have purified geometry, although Archimedes later mixed it with practical things.

Euclid's books on geometry, the *Elements,* began to circulate roughly around 250 BCE. The earliest extant fragments and copies of these books specify no author. The earliest historical trace of a Euclid who worked on geometry is a brief critical mention in a work by the geometer Apollonius, around 185 BCE.[1] Euclid was later mentioned, rarely and briefly, by Cicero and a few others.[2]

I don't know where Euclid was born or where he lived, but many writers say that he lived in Alexandria, Egypt. This popular guess evolved in the following way. At around 185 BCE, Apollonius claimed to have once spent some time in Alexandria. He also briefly mentioned Euclid, in a separate passage, making no reference to where or when he lived. Five centuries later, around 320 CE, Pappus of Alexandria claimed that Apollonius "spent a long time with the pupils of Euclid at Alexandria."[3] But note, Pappus did not say that Euclid himself lived in Alexandria. Next, around 460 CE, Proclus claimed that the Egyptian king Ptolemy I "once asked Euclid if there was not a shorter road to geometry than through the *Elements,* and Euclid replied that there is no Royal Road to geometry."[4] It's a great story, but there is no good reason to believe that Proclus really echoed events from more than seven centuries prior, when Ptolemy reigned in Alexandria. Nevertheless, knowing nearly nothing more about the mysterious Euclid, most writers and historians echo an old guess: that Euclid lived in Alexandria. And they often invent additional details: *when* he was born, that he was a student of Plato, that he was summoned to work for King Ptolemy, that he worked at the Library of Alexandria, and so on. At least there is fair evidence that Euclid was really a geometer, in contrast to Pythagoras.

In any case, Proclus was a pagan theologian and a mystic admirer of Plato. Accordingly, Proclus claimed that "Euclid belonged to the persuasion of Plato and was at home in this philosophy."[5] The conjecture is often illustrated by the following similarity. Plato had speculated that the five regular solids were the building blocks of the universe. And Euclid's *Elements* ends with the construction of these same five solids. But really, this does not mean that Euclid admired Plato, because over the centuries many people, before and after Plato, built or discussed these geometric figures. Still, did ancient geometers echo Plato's beliefs?

The *Elements* elegantly conveys a wealth of geometrical knowledge, systematically organized as propositions proven on the basis of basic principles. Interestingly, the *Elements* proceeds without much reference to physical matters. The propositions and proofs lack expressions that explicitly refer to motion. Geometric figures are constructed but formulated mostly as expressions that do not require any movement of figures in space or relative to one another. One exception is the fourth "common notion." It states: "Things that coincide with one another are equal to one another." This rule could be used to show that figures can be moved so that they overlap, and if their parts match exactly then they are equal. Historian Thomas Heath commented that Euclid's words left "no room for doubt that he regarded one figure as actually *moved* and *placed upon* the other."[6] Such displacements of figures were not specified explicitly in many of the demonstrations, but the fourth common notion was used in proposition 4 of the *Elements,* book 1, to move a triangle to superimpose it onto another. Afterward, that proposition is repeatedly used, explicitly or implicitly, and therefore Heath argued that the procedure of moving figures was fundamental.

Still, Heath inferred that Euclid probably preferred not to move figures, because apparently Euclid chose not to move figures in some propositions where that would have sufficed. Therefore, Heath conjectured that the fourth common notion was not originally in Euclid's *Elements* but had been added later by some anonymous editor. After all, this "common notion" is not mentioned by name in the geometric proofs in the *Elements.* Moreover, Heath himself disliked the idea of moving figures: "The method of superposition, depending on motion without deformation, is only of use as a practical test; it has nothing to do with the theory of geometry." Heath alluded to arguments by Arthur Schopenhauer and Bertrand Russell to undermine the idea that motion plays any fundamental role in geometry.[7]

According to Proclus, the Pythagoreans divided "the mathematics of magnitude" into two parts: geometry, the study of magnitude at rest, and "spherics," the study of moving magnitudes. Yet Proclus disagreed: "Magnitudes, figures and their boundaries, and the ratios found in them, as well as their properties, their various positions and motions—these are what geometry studies."[8]

Some great geometers used ideas of motion. An important example is Archimedes, in the third century BCE. This famous mathematician and engineer lived in Syracuse, Sicily, then part of Greater Greece. In his work "On Spirals," Archimedes described a spiral by a moving line: "If a straight line of which one extremity remains fixed be made to revolve at a uniform rate in a plane until it returns to the position from which it started, and if, at the same time as the straight line revolves, a point move at a uniform rate along the straight line, starting from the fixed extremity, the point will describe a spiral in the plane."[9] This account sounds physical and visual, but it may just be imagined. In the same text, Archimedes expressed the first two (of twenty-eight) propositions in terms of moving points, and he also established proportionality between described lengths and time intervals.

But most of the other propositions lack explicit references to motions in time. Likewise, many of the other geometrical works by Archimedes, on spheres and cylinders, conoids and spheroids, as well as a book of lemmas (usually attributed to Archimedes), followed the traditional style. Archimedes became famous, for example, for finding that $22/7 > \pi > 223/71$, for creating a numbering system to try to calculate how many grains of sand can fit in the universe, and for comparing a sphere to a cylinder.

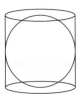

Figure 4.1. Archimedes showed that the ratio of the cylinder to the inscribed sphere is 3:2.

According to Plutarch, Archimedes was quite impressed by discovering that the sphere is two-thirds of the cylinder: "His discoveries were numerous and admirable; but he is said to have requested his friends and relations that, when he was dead, they would place over his tomb a sphere containing a cylinder, inscribing it with the ratio which the containing solid bears to the contained."[10]

In mathematics, to prove that the volume of a sphere is two-thirds that of a cylinder, we must carry out multiple geometric operations. But in practice, for a quick result, just submerge a ball into a can full of water. The works of Archimedes show other ways in which physical matters are related to geometry. His works "On Floating Bodies" and "The Equilibrium of Planes" blend geometrical and physical elements. They deal with questions about fluids, weight, volume, centers of gravity, levers, and balance, while analyzing all these questions almost without reference to physical causes, focusing on geometric relations, so that these physical matters seem to depend on pure geometry. Still, these works actually depended on experiments.

For example, in one work Archimedes first postulated that "equal weights at equal distances are in equilibrium, and equal weights at unequal distances are not in equilibrium but incline towards the weight which is at the greater distance."[11] Geometrically, we might think that if equal magnitudes lie equally far from a point, they must balance simply because of symmetry. But in 1883 physicist Ernst Mach pointed out that "we forget, in this, that a great multitude of negative and positive experiences is implicitly contained in our assumption," for example, that the figures' colors or shapes, or positions relative to other things and the observer, do not affect equilibrium.[12] Weights and relative distances have been selected as the only relevant factors for equilibrium. In Archimedes's writings, the abstract propositions correspond to physical facts.

In 1906, Johan L. Heiberg discovered a previously lost work of Archimedes. Heiberg was examining a Byzantine prayer book written in Greek on goatskin and kept at the "daughter house" (in Istanbul) of the Church of the Holy Sepulchre in the old walled city of Jerusalem. To make this prayer book (completed in 1229 CE, presumably in Constantinople), scribes had first scraped off old writings from the parchment folios, almost erasing the manuscript copies of seven treatises by Archimedes, copied sometime between 950 to 1000 CE. Heiberg discovered that one of these treatises was a work previously lost in its entirety. It became known as "The Method," because it explains how to discover certain propositions in geometry.[13] Interestingly, it shows that Archimedes used physical, mechanical experiments.

Archimedes wrote the text as a long letter to Eratosthenes of Cyrene, at Alexandria, to relate the mechanical procedures by which Archimedes had discovered various geometric propositions and solutions. He had written a similar procedure in his "Quadrature of the Parabola," which he sent to Dositheus of Pelusium, a mathematician at Alexandria, when a mutual friend died. Archimedes explained that although geometers had tried and failed to find a rectilinear area that is equal to the area of a circle, he had at least discovered a way to find something similar: a rectilinear area that equals a parabola. Archimedes had shown that the area of a parabola is four-thirds that of a triangle that has the same base and equal height. About this theorem, he stated, "I first discovered [it] by means of mechanics and then exhibited [it] by means of geometry." To do so, he first constructed triangles and parabolas out of some material and then suspended them from a balance. By cutting off pieces of the figures, he compared them. Using a particular fulcrum point, he showed how each line (or strip) in a triangle balances a line in a parabolic segment. Therefore, the sums of all these lines balance: the whole triangle balances the whole parabolic segment.

Archimedes's procedures also include abstractions from what is physically possible. For a century, since Heiberg's discovery, mathematicians and historians had believed that Archimedes had entirely avoided the concept of actual infinity, instead using only a notion of potential infinity, that is, that a figure can be divided into as many segments as necessary. Moreover, for centuries many writers had claimed that Greek mathematicians did not use the concept of actual infinity. But in 2001, following a new meticulous analysis of the deteriorated and moldy goatskin parchment that was by then more than a thousand years old, paleographers made a striking discovery.[14] A dozen lines of previously undeciphered and faded Greek text, in proposition 14 of Archimedes's "Method," seem to refer to actual infinity. Archimedes assumed that the number of lines inside a rectangle is equal to the number of triangles inside a prism. He wrote about the infinite "equal multitude" of such lines and triangles.[15]

In addition to "The Method," the faded old copies of Archimedes's works included a single page of another interesting work, previously lost almost in its entirety. Historians have argued that it is related to a mathematical puzzle whose name is translated sometimes as "the stomach ache." The puzzle is about taking a square, dividing it into fourteen pieces in a predefined pattern, shuffling the pieces, and then trying to reconstruct the square. But the single extant page by Archimedes was much too deteriorated, nearly indecipherable, to figure out what he contributed to the problem. Finally, new techniques for visualizing texts were applied to it, and historians now think that the text involved the question of figuring out how many possible ways the puzzle can be solved, that is, that this early text pertained to the field that many centuries later became known as combinatorics.[16]

The most famous anecdote about Archimedes is the story about King Hiero's crown. Book 9 of Vitruvius's *Ten Books on Architecture* (ca. 15 BCE) provides the following account:

> Indeed, although there were many wonderful and various discoveries of Archimedes, from all of these, one in particular, which I shall explain, seems to have been worked out with boundless ingenuity. For Hiero of Syracuse, having been enormously elevated by his royal power and successful accomplishments, since he had decreed that a golden votive crown should be dedicated to the immortal gods in a certain temple, he decided that it should be built at a workman's wage, and weighed the gold on a scale for the contractor. At this time, a man presented the work carefully performed by his hand on behalf of the king, and he seemed to have provided the weight of the crown according to a scale.
>
> After evidence was brought forth that some amount of silver had been mixed into the work of the crown, with the gold taken out, Hiero, angry that he had been insulted but not discovering the method by which he might detect the theft, asked Archimedes to undertake for himself the investigation into the matter. Then, this man, when he had the matter under consideration, came by chance to the bathhouse, and there when he went down into the pool, he noticed that as much of his own body sank into it, so much water flowed out of the bath. And so, since he had demonstrated the method of the explanation of the matter, he did not delay, but moved with joy he jumped out of the pool, and going home naked, indicated with a clear voice that he had quite accurately discovered that which he sought; for, running, he was shouting at the same time, in Greek, "I found it! I found it!" ["Eureka! Eureka!"]
>
> Then indeed, because of this initial discovery, he is said to have

built two masses each of equal weight as the crown: one from gold, and the other from silver. When he had done this, he filled a vase full to the uppermost rim with water, into which he placed the mass of silver. As much as this mass sank down into the vase, so much water flowed out. So, with the mass taken out, he filled in with a measure the amount by which the diminishment had been made, so that with the same method by which he had proceeded, he might level [the water] at the rim. And so from this [method], he discovered how much corresponded to a certain weight of silver [and] a certain measure of water.

When he had put this to the test, he then similarly placed a mass of gold in the full vase, and, with it taken out, with the measure added by the same method, he discovered from the amount of water that it was not the same: that the mass of gold was smaller by a smaller quantity in the amount of matter in the same weight as the mass of silver. Afterwards, with the crown itself placed down into the same water—with the vase having been refilled—he found that more water had flowed into the crown than into the mass of gold of the same weight, and so, because of the fact that there was more water in the crown than in the mass, he rationally detected the mixture of silver into the gold and the manifest fraud of the contractor.[17]

When a lump of silver and a lump of gold have the same weight, however, they have different sizes. Reportedly Archimedes therefore realized that if the crown had silver in it, it would displace a different amount of water than would a pure gold crown of the same weight.

Did these events really happen? At first, the story seems plausible because Archimedes had written works both on floating bodies and on the balance of weights. However, analyses of the experiment described cast doubt on the story. A cube of pure gold weighing 1,000 grams occupies nearly 52 cubic centimeters. If we immerse it in a cylinder full of water, measuring 10 centimeters in radius, we can divide the volume of the gold by the aperture of the container:

$$\frac{\text{volume of gold}}{\text{container aperture}} = \frac{52\text{cm}^3}{\pi r^2 \text{cm}^2} = \frac{52\text{cm}^3}{314\text{cm}^2} = 0.17\text{cm}$$

That's how much the water would rise if a crown or wreath of pure gold, weighing 1,000 grams, were immersed in it. By contrast, a cube of pure silver weighing 500 grams occupies nearly 48 cubic centimeters. So if half of a crown weighing 1,000 grams is made of gold and the other half is made of silver, then we can calculate that this composite crown, placed by itself in the previous container full of water, would raise the water by 0.24 cm. The difference between how much water is displaced by the composite crown as is displaced by one of pure gold is 0.24 − 0.17 = 0.07 cm, less than a single millimeter!

For centuries, therefore, commentators have argued that the difference in water displacement is too small to be accurately measured, because the surface of water is not completely flat, air bubbles can attach to the wreath, water can cling to each object when it is removed, and so on. This arrangement assumes that the container full of water is a cylinder with a large opening. What if we use a smaller aperture? If we use a vertical container, say—let its aperture be 2 centimeters × 10 centimeters—and we immerse the crown sideways, to increase how much the water rises, it still would rise by just 2 millimeters, difficult to measure given the various sources of error.

Since the procedure described by Vitruvius does not work in practice, with the accuracy available in the time of Archimedes, writers have devised various other procedures that would work. In particular, if we hang a crown from one side of a balance, and hang a lump of pure gold weighing the same from the other side, at equal distances from the fulcrum, the two weights will balance. But if we then submerge the entire balance, with the two suspended objects, wholly into water, the two will no longer balance if they are not both made of pure gold. Water exerts less resistance against heavy, small sinking objects than against larger objects, or as physicists say, water exerts more buoyancy on less dense objects.

The story about Archimedes and the crown was echoed by Plutarch, by Proclus, and by many later writers. In 1586, twenty-two-year-old Galileo Galilei claimed that the procedure by which Archimedes had solved the puzzle remained unknown. It's interesting to read how Galileo thought about the past and how he construed the evolution of this story:

> Archimedes discovered the blacksmith's swindle in the golden crown
> of Hiero, but I think that until now the procedure that this great
> man must have used in this discovery has remained unknown. . . . I
> well believe that, as the story spread that Archimedes had discovered
> the fraud by means of water, some writer of that time recorded that
> fact; and that the same author, to add something to the brief tale
> he had heard, said that Archimedes used water in the way that was

commonly known. But I know that this procedure was entirely false and lacking the accuracy required in mathematical matters, which made me think several times how, by means of water, one could exactly ascertain the mixture of two metals, and finally, after having diligently reviewed what Archimedes demonstrates in his books "On things that float in water" and that "On things that weigh the same," a method came to my mind which exquisitely solves our puzzle: which method I believe is the same used by Archimedes, because, besides being very exact, depends moreover on the very demonstrations discovered by Archimedes himself.[18]

Galileo's account is typical, reminiscent of how people assessed stories about Pythagoras. Having read an old story, someone realizes that part of it is false. But it's a good story, so the reader assumes that the rest of the story is true and therefore proposes an alternative scenario that would fix the defect in the actual story. The conjecture seems very plausible; therefore its author presumes to have discovered the truth. Thus Galileo said he knew how Archimedes solved the puzzle. Personally, I don't know if Archimedes solved a golden puzzle for a king, but regardless of whether it happened, the puzzle itself is really great.

Another famous story about Archimedes was told by the Roman historian Plutarch, at around 100 CE. He wrote that Archimedes had once told King Hiero of Syracuse that given a force, any weight could be moved and that he boasted

that if there were another Earth, by standing on it he could move our Earth. Hiero being struck with amazement at this, and entreating him to show this claim by actual experiment, to show some great weight moved by a small engine, Archimedes turned to a cargo ship from the king's arsenal, which could not be drawn out of the dock without great labor and many men; and, loading her with many passengers and a full freight, he sat far off, with no great endeavor, but only holding the head of the pulley in his hand and drawing the cords by degrees, he drew the ship in a straight line, as smoothly and evenly as if she had been in the sea.[19]

Likewise, the story about Archimedes's death is great. Its longest extant early version is also by Plutarch. Marcus Claudius Marcellus was a prestigious military leader of the Roman Republic. In 214 BCE, Marcellus attacked the kingdom of Syracuse, centered on the city of Syracuse, a fortified city by the sea that formerly had been a Greek city and was still filled with beautiful works of Greek art and architecture. Archimedes lived in Syracuse, and Marcellus struggled to penetrate the city, failing repeatedly, partly because Archimedes had prepared machines of war to defend the city walls.

The battle transpired for two years. Three centuries later, Plutarch described it as follows. The Roman army surrounded Syracuse by land and sea. From warships, they attacked the city walls: "With sixty galleys, each with five rows of oars, carrying all kinds of weapons and missiles, and a huge bridge of planks laid upon eight ships chained together, upon which was carried the engine to cast stones and darts, they assaulted the walls," said Plutarch, "all of which, however, seemed to be but trifles for Archimedes and his machines." King Hiero had employed Archimedes to build various engines of war, especially to defend the city during a siege. Plutarch wrote:

> Therefore, when the Romans attacked the walls in two places at once, fear and consternation stupefied the Syracusans, thinking that nothing could withstand that violence and those forces. But when Archimedes began to ply his engines, he at once shot against the land forces all sorts of missile weapons, and immense masses of stone that came down with incredible noise and violence; against which no man could stand; for they knocked down those upon whom they fell in heaps, breaking all their ranks and files. In the meantime huge poles thrust out from the walls over the ships sunk some by the great weights which they let down from on high upon them; others they lifted up into the air by an iron hand or beak like a crane's beak and, when they had drawn them up by the prow, and set them on end upon the poop, they plunged them to the bottom of the sea; or else the ships, drawn by engines within, and whirled about, were dashed against steep rocks that stood jutting out under the walls, with great destruction of the soldiers that were aboard them. A ship was frequently lifted up to a great height in the air (a dreadful thing to behold), and was rolled to and fro, and kept swinging, until the mariners were all thrown out, when at length it was dashed against the rocks, or let fall. At the engine that Marcellus brought upon the bridge of ships, which was called Sambuca, from some resemblance it had to a musical instrument, while it was as yet approaching the wall, there was discharged a piece of rock of ten talents weight, then a

second and a third, which, striking upon it with immense force and a noise like thunder, broke all its foundation to pieces, shook out all its fastenings, and completely dislodged it from the bridge.[20]

Marcellus and his fleet retreated in despair. They attacked again in the darkness of night, trying to scale the city walls, but a shower of stones, arrows, darts, and other missiles again forced them to withdraw. The machines of Archimedes were set behind the walls, "whence the Romans, seeing that indefinite mischief overwhelmed them from no visible means, began to think they were fighting with the gods." Marcellus allegedly complained that Archimedes outmatched "the hundred-handed giants of mythology."

While the Romans became increasingly terrified of Archimedes, who seemed to make Syracuse invincible, they nonetheless defeated other Greek cities and towns, killing thousands of people. And eventually, Marcellus noticed a tower in the walls of Syracuse that seemed vulnerable. The Romans built suitable ladders, and one night, while the Syracusans were celebrating a feast to the goddess Diana, busy drinking wine, the Roman soldiers infiltrated the tower. Before dawn, and before the citizens noticed, they filled the tower and the surrounding wall with soldiers. Then combat broke out, and the trumpets of Marcellus frightened many Syracusans, who promptly fled. Triumphant, Marcellus yet feared that his soldiers would plunder and destroy parts of the beautiful city. Plutarch said that Marcellus therefore ordered the Romans to exercise restraint: that the buildings should not be burned; that although money and slaves could be seized, no free person should be abused, killed, or enslaved. "Despite this moderation, he still felt pity for the city," Plutarch wrote,

> for soon they plundered the other parts of the city, which were taken by treachery; leaving nothing untouched but the king's money, which was brought into the public treasury. But the suffering of Archimedes most afflicted Marcellus. For he then happened to be gazing at a diagram, for having given at once his mind and his sight to the problem, he did not notice the invasion of the Romans or the sack of the city, and suddenly a soldier lunged upon him, ordering him to go to Marcellus, he was unwilling before both finishing the problem and establishing its demonstration. So the man became enraged, drew his sword, killed him. Others say, however, that when a Roman attacked him and was about to kill him, sword in hand;

that he, seeing him, pleaded and begged him to wait a short while, so that he might not leave his work unfinished and unproven, but who, regardless, killed him. And there is a third story, that when he was carrying some of his mathematical instruments, dials and spheres and angles, by which the size of the Sun could be measured by sight, some soldiers saw him and thinking he carried gold in the box, killed him. Yet, all agree that Marcellus was aggrieved and banished the murderer as if cursed, and honored [Archimedes's] relatives, having sought them out.[21]

Plutarch's account of Archimedes's last years is not the earliest, but it is the longest. If Archimedes really died when the siege of Syracuse ended, then he died in 212 BCE. The earliest extant account of his death is from over a century later, when Cicero briefly reported around 70 BCE that Archimedes was killed despite Marcellus's interest in him. Table 4 shows various ancient and very old accounts of how Archimedes died.[22] These accounts convey various ideas about Archimedes: they repeatedly claim that he was rather absentminded or absorbed in geometry, but some also claim that he begged for respect for his diagram; that he was disobedient or asked for time to finish his diagram; that he was murdered for gold; or that he resisted capture, bravely challenging the Roman soldier.

Two centuries after his death, or sooner, alleged quotations arose of Archimedes's final words: "I beg you, don't disturb this." And nearly two thousand years after his death, this phrase became attributed to him: "Do not disturb my circles." It appeared in a work of 1621, as a Latin phrase: "Noli turbare meos circulos."[23]

We can compare the early evolution of the story of Archimedes's death with more recent accounts. Table 5 shows various accounts from the last hundred years, written by popular or famous writers.[24] Accounts by historians, by contrast, have been more reliable, at least in echoing Plutarch's words. Interestingly, the recent accounts add details that are absent in the ancient sources. For example, that the Roman soldier stepped on the geometrical diagram, that Archimedes was finishing a proof, that Archimedes was in a courtyard or in a beach, or that he was not used to taking orders.

Some accounts overtly say that Archimedes died as a "martyr" for mathematics.[25] They portray old Archimedes as a noble hero devoted to an admirable cause; he faced opposition bravely, risked his life, and was ruthlessly killed. In ancient Christianity, martyrs were those who testified to the Word

Table 4. Ancient and Medieval Accounts of the Death of Archimedes

ca. 212 BCE	Archimedes died.	
ca. 70 BCE?	Cicero	"Marcellus. In truth he is said to have diligently sought the renowned Archimedes, a man of the highest talent and learning, and upon hearing that he was killed, became greatly disturbed."
45 BCE	Cicero	"Consider what passion for study was in Archimedes, who while attentively drawing in the dust, did not realize that his fatherland had been captured?"
ca. 15 BCE?	Titus Livius	"Tradition reports that Archimedes, in such great tumult, while the city was captured by terror of the plundering soldiers, he was intent on figures that he drew in the dust, when a soldier killed him not knowing him; it distressed Marcellus, who gave him a proper burial."
ca. 30 CE?	Valerius Maximus	"But he, with mind and eyes fixed on drawing figures on the ground, a soldier broke into his house to plunder with sword drawn over his head, demanding who he was, but because of his deep desire to investigate what he sought, he did not manage to say his name, but with his hands protected the dust and said, 'I beg you, don't disturb this!' and then, as if unaware of his victor's demand, his blood confused the lines of his art."
ca. 100 CE	Plutarch	"But the suffering of Archimedes most afflicted Marcellus. For he then happened to be gazing at a diagram, for having given at once his mind and his sight to the problem, he did not notice the invasion of the Romans or the sack of the city, and suddenly a soldier lunged upon him, ordering him to go to Marcellus, he was unwilling before both finishing the problem and establishing its demonstration. So the man became enraged, drew his sword, killed him."
ca. 100 CE	Plutarch	"Others say, however, that when a Roman attacked him and was about to kill him, sword in hand; that he, seeing him, pleaded and begged him to wait a short while, so that he might not leave his work unfinished and unproven, but who, regardless, killed him."
ca. 100 CE	Plutarch	"And there is a third story, that when he was carrying some of his mathematical instruments, dials and spheres and angles, by which the size of the Sun could be measured by sight, some soldiers saw him and thinking he carried gold in the box, killed him."
ca. 1150	John Tzetzes	"He was hunched over, drawing a mechanical diagram, but a Roman seized him, taking him away as prisoner. But he, being wholly intent on the diagram at that time, and not knowing who was dragging him, said to that man: 'Step back, man, from my diagram.' And as he was manhandled, turning around and realizing that it was a Roman, he shouted: 'Somebody give me one of my machines!' But the Roman, scared, killed him right then, a man feeble and old, but ingenious in his works."
ca. 1150	John Zonaras	"The Romans became the rulers of these regions, killing many others, and Archimedes. For drawing a diagram and hearing that the enemy was attacking, he said 'Come at my head, but not at my line!' As an enemy confronted him, he was a bit perturbed, saying, 'Step back, man, from my line,' so provoked, the man struck him down."

Table 5. Relatively Recent Accounts of How Archimedes Died, by Popular Authors

ca. 212 BCE?	Archimedes died.	
1937	E. T. Bell	"His first intimation that the city had been taken by theft was by the shadow of a Roman soldier falling across his diagram in the dust. According to one account the soldier had stepped on the diagram, angering Archimedes to exclaim sharply, 'Don't disturb my circles!'"
1971	Petr Beckmann	"Suddenly a soldier came up to him and bade him follow to Marcellus, but he would not go until he had finished the problem and worked it out to the proof. 'Do not touch my circles!' said the thinker to the thug. Thereupon the thug became enraged, drew his sword and slew the thinker."
1972	Morris Kline	"While drawing mathematical figures in the sand, Archimedes was challenged by one of the Roman soldiers who had just taken the city. Story has it that Archimedes was so lost in thought that he did not hear the challenge of the Roman soldier. The soldier thereupon killed him."
1988	Paul Hoffman	"When one of Marcellus' men found Archimedes in a courtyard, drawing geometric figures in the sand, he disobeyed his orders and drew his sword. 'Before you kill me, my friend,' Archimedes pleaded, 'pray let me finish my circle.' The soldier did not wait. As Archimedes lay dying, he said, 'They've taken away my body, but I shall take away my mind.'"
2003	Timothy Ferris	"Archimedes was absorbed in calculations when a Roman soldier approached and addressed him in an imperative tone. Archimedes was seventy-five years old and no fighter, but he was also one of the freest men who ever lived, and unaccustomed to taking orders. Drawing geometrical diagrams in the sand, Archimedes waved the soldier aside, or told him to go away, or otherwise dismissed him, and the angry man cut him down."
2005	Stephen Hawking	"Legend records that the Roman soldier found Archimedes drawing figures in the sand. The soldier commanded Archimedes to stop what he was doing and leave immediately. Archimedes asked for more time to work out a problem in the sand. Enraged, the soldier ruined Archimedes's figures in the sand and ran him through with his sword!"
2007	Eli Maor	"Marcellus ordered his troops not to harm the great scientist. A soldier found the old sage on the beach, hunched over a figure he had drawn in the sand. Ignoring the soldier's order to stand up, Archimedes was slain."
2010	Amir Alexander	"He was killed when, oblivious to the sack of the city, he asked an ignorant Roman soldier to wait while he worked out a geometrical problem."

of God, the Gospels, even at a painful price, sometimes death. Thus martyrs followed the example of Jesus Christ in their willingness to sacrifice their lives for religious truth. In the story of Archimedes, he sacrifices himself in service of geometry. His example inspires us to value mathematics.

We saw that the story that the cult of Pythagoras murdered Hippasus for discovering irrationality is fiction. Yet there are other stories that function as fair stories about martyrs in the history of mathematics. Most historians agree that plenty of evidence shows that old Socrates was executed by the Athenians. Socrates, a proponent of the importance of math in education, was killed by the Athenians for teaching young men to think critically, to value reason above local religious doctrine. Likewise, most historians believe that the great geometer Archimedes was killed by a Roman soldier in the siege of Syracuse, and they at least echo the tale that at that moment he was busy at work in geometry. I do not know if these two stories are true, but they are worth studying nevertheless.

I find it fascinating that the existence of Archimedes is evinced clearly by the last trace of his death. At some point, the Roman statesman Cicero searched for the tomb of old Archimedes, and he found it. This is often noted, but his actual report is rarely quoted in histories of ancient mathematics. So here is Cicero's report, from around 45 BCE:

> I will tell you of a humble and obscure mathematician of the same city, called Archimedes, who lived many years after; whose tomb, overgrown with shrubs and briers, I discovered in my quæstorship, when the Syracusans knew nothing of it and even denied that any such thing remained; for I remembered some verses which I had been told were engraved on his monument, and they said that on the top of the tomb there was placed a sphere with a cylinder. When I had carefully examined all the monuments (for there are very many tombs at the gate Achradinæ), I found a small column standing out a little above the briers, with the figure of a sphere and a cylinder on it; whereupon I immediately said to the Syracusans—for there were some of their leaders there with me—that I thought that was what I was searching for. Several men came with scythes, cleared the way, and made an opening for us. When we could get at it, and came near the front base of it, I found the inscription, though the latter parts of all the verses were effaced almost half away. Thus one of the

noblest cities of Greece, which had been celebrated for its learning, had known nothing of the monument to its most ingenious citizen, if it had not been discovered to them by a native of Arpinum. But return to the subject from which I digressed. Who is acquainted with the Muses, that is, with liberal knowledge, or who deals at all in learning, who would not choose to be this mathematician rather than that tyrant? If we look at their lives and occupations, we find the mind of the one strengthened and improved, tracing the deductions of reason, amused with his own ingenuity, the sweetest food of the mind; the thoughts of the other engaged in continual murders and injuries, in constant fears by night and by day. Now imagine a Democritus, a Pythagoras, and an Anaxagoras; what kingdom, what riches would you prefer to their studies and amusements?[26]

5

GAUSS, GALOIS, AND THE GOLDEN RATIO

We tend to fit history into the forms of traditional stories about heroes, victims, and martyrs, struggle, success, and injustice. Hence we read: "From of old it has been the custom, and not in our time only, for vice to make war on virtue. Thus Pythagoras, with three hundred others, was burnt to death."[1] But really, we do not know just how Pythagoras died. Certain stories resonate: the martyr who was punished for speaking truth to power, the unassuming guy who did good against all odds, the wonder boy who cleverly solved the daunting problem, the unappreciated worker whose brilliant contributions were stolen, the forgotten pioneer who made discoveries ahead of his time, the genius who was a master of many fields, the disciple who rebelled against his master. We gradually mold our impressions to fit such patterns.

There is a widespread story that in 1786, a lazy schoolteacher in Germany gave his students the task of adding the first one hundred integers, just to keep them busy. An ordinary student might begin to write

$$1 + 2 = 3,$$
$$3 + 3 = 6,$$
$$6 + 4 = 10,$$
$$10 + 5 = 15,$$
$$15 + 6 = 21,$$

and students might continue adding for hours before they reached the final result, adding all one hundred numbers. But immediately, the youngest student, nine-year-old Carl Gauss, provided the answer by "folding" the series to add all of the numbers in pairs:

$$1 + 100 = 101, 2 + 99 = 101, 3 + 98 = 101, \ldots = 101 \times 50 = 5050$$

and the boy stunned the teacher by also giving the general formula for adding numbers 1 through n:

their sum = $n(n + 1) / 2$

The story is told by various historians and biographers of Gauss.[2] Again, this story is less than history, or more. Physicist and writer Tony Rothman remarks: "Rest assured that mathematicians are second to none in the uncritical acceptance of their own mythology."[3] But beyond acceptance, what is at stake is a creative process of interpolations.

This popular tale about Gauss is the product of decades of evolution, its history traced painstakingly by Brian Hayes.[4] He analyzed more than seventy versions of the story and traced their roots to a memorial tribute first published in 1856, the year after Gauss died. In that early version, Wolfgang Sartorius claimed that the old Gauss used to recall a childhood incident with relish: that at the beginning of a class, Mr. Büttner, a schoolteacher with a stick whip, gave a problem of summing an arithmetic series, but immediately Gauss solved it, threw his slate on the table, and said: "There it lies." The teacher waited until all students had finished counting, multiplying, and adding, occasionally looking at the youngest boy with pity. But when he checked all the results he found that only the young Gauss was correct, all others wrong.[5]

What Hayes found was that this and the other early versions of the story, from 1856 to 1933, did not include many of the key details featured in the now popular version. They did not state the particular series of numbers that Büttner asked the students to add, and they did not specify Gauss's solution either. Yet in 1937, the mathematician Eric Temple Bell purported a specific series of numbers: 81297, 81495, . . . , 100899, as Gauss's supposed puzzle.[6] Bell was president of the Mathematical Association of America, so he seemed to be a trustworthy authority. But he mixed fiction with history. From 1938 to the present, a few writers have parroted Bell's numbers.[7] Meanwhile, many other writers have given instead the simpler series 1, 2, 3, . . . , 100.[8] In 1966 another writer gave instead the series 0, 1, 2, 3, . . . , 100.[9] In 1984, another book gave the series 11, 14, 17, . . . , 26, instead.[10] Various other series also have been purported, in accounts dating from 1938 to the present, along with various solutions supposedly provided by the young boy Gauss.

Writers added details to the story: that the schoolteacher, Büttner, was a brute who thrashed and terrified, or that he was lazy and wanted to give busy work to the students, or that it was a punishment for the unruly class, or that it was just a mindless problem, that he did not know the answer beforehand or alternatively that he did know it. In 2001, three fifth-graders wrote their version of the story, and they too invented details: they stated that the teacher did not like math and did not believe that Gauss had really solved the problem, so he made the boy show the class how he had done it. The fictions invented by children were similar to those published in 1934 by Bell, a president of the Mathematical Association of America. Hayes remarked: "Am I being unfair in matching up Eric Temple Bell against three fifth-graders? Unfair to which party? Both offer interpretations that can't be supported by historical evidence, but Ryan, Jordan and Matthew are closer to the experience of classroom life." Hayes explained: "Tellers of a tale like this one seem to work under a special dispensation from the usual rules of history-writing. Authors who would not dare to alter a fact such as Gauss's place of birth or details of his mathematical proofs don't hesitate to embellish this anecdote, just to make it a better story. They pick and choose from the materials available to them, taking what they need and leaving the rest—and if nothing at hand suits the purpose, then they invent!"[11] Hayes found that what drives the evolution of this story is not merely an accumulation of errors of transmission, but instead authors' deliberate choices to improve the story, an urge to explain by adding details to polish the narrative.

A similar case of cumulative inventions led to the popular legend of Évariste Galois. On 30 May 1832, the twenty-year-old Galois died in a pistol duel; that much is true. But mathematicians and scientists used to believe, some still do, that on the previous night Galois had suddenly and feverishly formulated the foundations of group theory, working against time.

This myth was impelled, again, by Eric Temple Bell, who was also a writer of science fiction. Bell imagined Galois as a victim of unfair negligence and persecution, that rejections pushed him to leave mathematics to turn to revolutionary politics and a liaison with a "worthless girl" that led to his death. Bell relished the moral of the story: "In all the history of science, there is no completer example of the triumph of crass stupidity over untamable genius."[12] But Tony Rothman rightly detailed how Bell concocted this story by omitting facts, disregarding chronology, and worst of all freely interpolating fictions.[13] In actuality, in 1830 Galois had already published three of his four papers on group theory; his revolutionary political activities actually were simultaneous with his mathematical labors; and there is no evidence that the young lady he liked, a physician's daughter, was worthless or, as in later accounts, a prostitute. Other writers interpolated additional fictions, mostly to suit personal biases and guesswork. Physicist Leopold Infeld conjectured that Galois had been killed by a police spy.[14] Similarly, physicist Fred Hoyle claimed that Galois had been killed in a political duel.[15] This led Rothman to complain: "The purpose has been to show that something is curiously out of sync. Two highly respected physicists and an equally well-known mathematician, members of the professions which most loudly proclaim their devotion to Truth, have invented history."[16]

Regarding Bell, in particular, Rothman further commented: "As an inventor of fairy tales, one can enjoy Bell; as a biographer it is unclear how far one can forgive him. Surely all his mistakes did not result from a poor knowledge of French. No, I believe Bell saw his opportunity to create a legend." I refer readers to Rothman's excellent work for a detailed accounting of Bell's many mistaken claims. Since the early 1920s, aside from working on mathematics, Bell wrote poems, stories, and science fiction novels. He published his works of fiction under the pen name "John Taine." He wrote about atomic energy, intelligent apes, genetic insanity, time travel, radioactivity, a superhuman, dinosaurs, and more. At the time, his ideas were relatively original and well developed, and he described his skill as follows: "It may not be apparent in these books to the casual reader that there is creative thought in them. . . . barring ordinary human mistakes, my training enables me to write stories that, no matter how wild apparently, are logically selfconsistent. They do not fall apart; within their own data they hang together."[17] These same words might roughly be applied to his biographies of mathematicians. By the time Bell's *Men of Mathematics* was first published, in 1937, it was well known that he and John Taine were the same person.

Bell's biographer Constance Reid has noted that "his successors in the history of mathematics have exposed factual errors and exaggerations seemingly without end."[18] Bell strikingly failed at accuracy in many points of history. But nevertheless, his *Men of Mathematics* was an enormous success in other ways. It generated much interest in the history of mathematics, it inspired young students to become mathematicians, and it helped general readers to appreciate mathematics. Bell's portrayal of Galois, especially, impressed and moved readers because it conveyed a traditional story of struggle:

a talented young man fighting bravely against a conservative and mediocre establishment. Nevertheless, the story of Galois is fascinating, even without Bell's inventions and guesswork.

In 1828, the talented Galois took the entrance examination for the prestigious École Polytechnique a year early but failed. In May and June of 1829, Galois submitted to the Academy of Sciences two outstanding papers on the theory of equations. His beloved father was the mayor of their hometown, Bourg-la-Reine near Paris, but having been framed as the author of malicious epigrams, his father committed suicide by asphyxiation on 2 July 1829. A few days later, Galois again failed the examination at the Polytechnique and became increasingly embittered. Still, Galois finished school and enrolled at the École Préparatoire (later known as the École Normale). Finally, six months after Galois had submitted his two papers to the Academy of Sciences, the appointed reviewer, Augustin Louis Cauchy, did not present them to its members, first because he was "indisposed at home" and later for reasons that are unclear. Rothman conjectures that perhaps Cauchy suggested that Galois should instead submit his papers for a prize competition. In any case, Galois did submit a paper for a competition for a Grand Prize in Mathematics: he sent it to Jean Baptiste Fourier, the Academy's secretary of mathematics and physics. But on 16 May 1830, Fourier died, and Galois's entry was not found among Fourier's papers or in the hands of the judging committee members. It was strangely lost, so the prize was awarded to someone else.

Nevertheless, Galois soon published three papers, which later became known as the Galois theory. Meanwhile, in late July 1830, crowds of people in Paris rioted against King Charles X. They promptly organized a major rebellion, inspired by the bodies of martyrs, citizens killed by soldiers in the streets. On 28 July, during the uprising, many students at the École Normale wanted to join the fights in the city, including Galois, but the director of the École locked the students in. Galois was so annoyed that he tried to escape by climbing the walls but failed, thus missing the dangerous turmoil. The revolution was successful, the king abdicated, and the royalists fled, including Cauchy in September. A constitutional monarchy was then established, headed by a new king, Louis Philippe. Unsatisfied, Galois joined a secret republican group, the Society of Friends of the People, and when the director of the École criticized students in the newspaper, Galois submitted a letter criticizing him. Therefore, he was going to be expelled from the École, but he immediately quit and joined an artillery unit.

In January of 1831, Galois submitted a new paper to the Academy, at the invitation of Siméon Denis Poisson. Meanwhile, Galois's bitter attitude made him increasingly unpleasant. For example, in April 1831 the mathematician Sophie Germain, while she was very ill, wrote to a friend:

> Really, there is a doom in all that relates to mathematics, your preoccupation, that of Cauchy, the death of Mr. Fourier, to finish off his student Galois who despite his impertinence showed a happy disposition, has done so much that he has been expelled from the École Normale. He has no money, and his mother has very little. Having moved back to her home, he continued his habit of insults toward her, of which he gave you a sample after your best lecture at the Academy. The poor lady fled her house, leaving just enough for her son to live tolerably, and has been forced to place herself as a lady's companion [caregiver] in order to meet her needs. They say that he will go completely insane, and I believe it.[19]

On 9 May 1831, roughly two hundred republicans gathered at a banquet to celebrate the trial acquittal of nineteen republican officers who had been arrested for conspiring to overthrow the monarchy.[20] At that defiant gathering, the rowdy Galois stood up and apparently made a threatening comment about the new king while brandishing a dagger, causing a loud tumult. The next day, Galois was arrested at his mother's house, for allegedly threatening the king's life. He was jailed at Sainte-Pelagie prison. But soon, the jury acquitted Galois.

On 4 July, in the name of the Academy of Sciences, Poisson rejected Galois's paper. Meanwhile, Galois continued to behave as a political radical. On Bastille Day, 14 July 1831, he participated in a republican demonstration, wearing the outlawed uniform of the Artillery Guard and carrying his dagger, pistols, and a loaded rifle in the streets. For these reasons, he was arrested again and sentenced to six months in prison. While he was there, his cellmates one day pressured him to get drunk, though he hated liquor, and once intoxicated he suddenly attempted to kill himself with some weapon, but his cellmates stopped him.[21]

At some point, the twenty-year-old Galois heard that Poisson had rejected his mathematical paper. Poisson noted that he had just not understood it but that he expected the young author to clarify and expand it. To convey these circumstances, it helps to read some of what Galois himself then wrote. In particular, a dramatic record of how snubbed he felt is conveyed in a preface he wrote for his "Two Memoirs on Pure Analysis," in late 1831, while he was still in prison:

> First, the second page of this work is not burdened by the last names, first names, qualities, dignities, and praise of any greedy prince whose purse opens at the smoke of incense with the threat of closing when the incense holder is empty. Neither does one there see, in letters three times bigger than the text, a respectful tribute to some high official in the sciences, to a wise patron, such an indispensable (I would say inevitable) thing for a twenty-year-old who wants to write. I tell nobody that I owe to his advice or his encouragement all that is good in this work. I do not say so: because that would be lying. If I have addressed anything to the greats of the world or to the greats of science (and in these days, the distinction is imperceptible between those two kinds of people), I swear that it would not be thanks. I owe some of them for making the first of these two papers appear so late, others for my having written it all in jail, a site that one can hardly regard as a place for contemplation and where I often found myself dumbfounded at my own recklessness for not keeping my mouth shut toward my stupid Zoïles: and I think I can use the word Zoïle in all certainty and modesty, as my adversaries are low in my mind.[22]

The name "Zoïle" refers to an ancient Greek literary critic and Cynic philosopher who lived in the time of Aristotle, in the fourth century BCE. Zoilus was infamous for his bitter critiques of the writings of Homer. Centuries later, the architect Vitruvius despised Zoilus as a slanderer and wrote the following lines about him: "Various stories are told about his death, which was like that of one found guilty of parricide. Some writers have said that he was crucified by Philadelphus; others that he was stoned at Chios; others again that he was thrown alive upon a funeral pyre at Smyrna. Whichever of these forms of death befell him, it was a fitting punishment and his just due; for one who accuses men who cannot answer and show, face to face, what was the meaning of their writings, obviously deserves no other treatment."[23] Owing to Zoilus's *Homeric Questions,* his name became synonymous with harsh and malignant criticism, and he became known as the scourge and "whipper" of Homer. In *Don Quixote,* Cervantes called Zoilus "badmouthed."[24] Thus a proverb arose: "Every great poet has his Zoilus."[25] So when prominent academic Frenchmen disregarded the works of the impassioned young mathematician Évariste Galois, he felt that he had found his Zoiluses. So his preface continues: "It is not my topic to say how or why I was detained in prison, but I must say how manuscripts get lost so very often in the folders of Gentlemen Members of the Institute, though in truth, I cannot conceive such carelessness on the part of men who have on their conscience the death of Abel."[26]

Galois referred to the young and talented Norwegian mathematician Niels Henrik Abel. In 1823, Abel wrote a paper in French, titled "A General Representation of the Possibility to Integrate All Differential Formulas," and submitted it for publication, but it was lost while a Frenchman was reviewing it. Still, he next published various mathematical papers in Crelle's *Journal* (in Berlin), while living in Berlin and other cities, but in 1826 he moved to Paris, there finishing what he regarded as his most important work and submitting it to the prestigious French Academy of Sciences. But they neglected his paper and did not publish it. In Paris, Abel lived in relative poverty and contracted tuberculosis; having failed to publish in French, he returned, unsuccessful, to Berlin and Norway. He died in 1829, age twenty-six. Thus Galois alluded to the injustices suffered by Abel:

> That is enough, for I do not want to compare myself to that illustrious mathematician; it will suffice to say that my paper on the theory of equations was effectively submitted to the Academy of Sciences in the month of February 1830, that parts of it were sent in 1829, that any trace was not followed up on, and that it has been impossible to recover the manuscripts. In this genre there are very bizarre anecdotes, but it would be graceless to recount them, because no similar accident, except the loss of my manuscripts, concerns me. Happy voyager, my bad demeanor has saved me from the jaws of wolves. I have already said plenty to make the reader understand why, what was once my goodwill, it has been absolutely impossible for me to decorate or disfigure, as you will, my work with a dedication. . . .
>
> . . . The first paper is not virginal from a master's view: an extract sent in 1831 to the Academy of Sciences was examined by Mr. Poisson, who in session reported having not understood it at all. That, to our eyes, fascinated by the author's self-love, proved simply that Mr. Poisson did not want to or was not able to understand, but it certainly proves in the eyes of the public that my book means nothing.
>
> All of this therefore leads me to think that in the academic world

the work that I submit to the public will be received with the smile of pity; that the more indulgent will condemn me as confused; and that for some time I will be compared to Wronski or to those tireless men who every year find a new solution for squaring the circle. Above all I will have to bear the crazy laughter of the Gentlemen Examiners of candidates at the École Polytechnique (who by the way I am surprised not to see each a chair in the Academy of Sciences, because their place certainly is not for posterity), who, having the tendency to monopolize the printing of mathematics books, will not learn without being formalized that a young man twice rejected by them also has the desire to write, not didactic books certainly, but books of theory. All of the above I have said to prove that it is knowingly that I expose myself to the laughter of fools.[27]

These are the words of a bitter young man; it might be tempting to make him into a saint, an innocent victim of unfair and narrow-minded men. But as mentioned, there is evidence that Galois himself was a troublemaker.

By March 1832, he was ill and was therefore transferred out of prison, to a sanatorium. In May, Galois received a letter from a young woman. He soon copied it roughly on the backside of a letter to a friend. It read: "Break off this affair, I beg you. I don't have enough will to continue a conversation on this topic, but I will try to have enough to converse with you as I did before anything happened. Here, Mr. le [illegible] . . . had . . . who you should . . . but to me and no longer think about things that do not [illegible] exist and that will never exist."[28] By inspecting this manuscript with a magnifying glass, one researcher managed to decipher the name of a woman beneath a few ink deletions by Galois: "Stéphanie Dumotel."[29] She was the daughter of a fine physician at the sanatorium house where Galois stayed in the spring of 1832. In several manuscripts, Évariste included drawings of the letter "E," accompanied by the letter "S," and in other places "St" or "Ste."[30] Furthermore, he also copied the following letter, apparently from Stéphanie:

I took your advice, and I've thought about . . . what has happened . . . under whichever denomination that can be [illegible] to establish between us. The rest Mr. . . . be assured that doubtless there would never have been any gain; you suppose wrongly, and your regrets are wrongly based. True friendship hardly exists except between persons of the same sex. Above all . . . of friends. . . . Without doubt the *void*

that . . . the absence of all feeling of that kind . . . [illegible] trust . .
. but has been very [illegible] . . . you have seen me sad . . . asked me
the cause; I answered you that I have sorrows, which one has made
me endure. I thought that you would take that as anyone in front
of whom one lets a word come out for [illegible] . . . one is not . . .
the serenity of my ideas gives me the freedom to judge with much
reflection the people whom I see habitually; that was what made me
rarely regret my having been *mistaken* or allowed them to influence
me. I do not share your opinion about the [illegible] . . . more that the
[illegible] . . . require nor that . . . thank you sincerely for all of *them*
who you would well bring down in my favor.[31]

What was she writing about? It remains unclear, although at least it seems
that she had in some sense rejected Galois. In late May, in a letter to his best
friend, Galois said that he had no more happiness, that he hated the world,
that he was hopeless and "disenchanted with everything."[32] On the night of
29 May, he wrote letters to his republican friends, as if he expected to die the
next day. In one he wrote:

> I beg the patriots, my friends, not to reproach me for dying for
> something other than my country. I die victim of an awful coquette
> and of two dupes of that coquette. It is in a miserable mess that my
> life ends. Oh! why die for such a small thing, die for something so
> despicable? I call on heaven to witness that it was under constraint
> and force that I have yielded to a provocation that I've avoided by
> every means. I regret having said the grim truth to men so hardly
> able to grasp it in cold blood. But still I told the truth. I go to the
> grave with a conscience free of lies, clean of patriot blood. Goodbye! I
> did good with life for the public good. Forgive those who have killed
> me, they are of good faith.[33]

Likewise, that night Galois wrote the following letter to two good republican
friends:

> My good friends,
> I was provoked by two patriots. . . . It has been impossible for me to
> refuse. I ask you to forgive me for not having warned either one of
> you. But my opponents have put me *on my honor* not to warn any
> patriot. Your task is quite simple: prove that I fought against my will,

that is to say, after having exhausted all means of accommodation, and say whether I am capable of lying, to lie even for such a small thing, which is what it was about. Hold onto my memory, since fate has not given me enough life so that the country may know my name. I die your friend.[34]

Moreover, that night Galois wrote a letter to his best friend summarizing his findings in mathematics over the last four years: researching the solvability of equations by radicals and describing the transformations possible on an equation that is not solvable by radicals, thus helping to create the field that became known as group theory. Galois added a few comments in the margin of the work that Poisson had rejected. Next to one theorem, he wrote: "There are a few things left to be completed in this proof. I have not the time." He ended his letter as follows: "It is greatly in my interest to not be mistaken such that someone suspects me of having enounced theorems of which I don't have the complete proof. You will publicly beg Jacobi or Gauss to give their judgment, not on the truth, but on the importance of these theorems. After that, there will be, I hope, some men who will find benefit in deciphering all this mess. I embrace you with effusion."[35]

Early in the morning, on 30 May, Galois and another young man fired pistols at each other. Galois was seriously injured and was taken to a hospital. His brother Alfred saw him there, and reportedly Évariste told him, "Don't cry, I need all my courage to die at twenty." And shortly thereafter he died. A few days later, a newspaper reported:

> A dreadful duel yesterday has deprived the exact sciences of a young man who gave the highest expectations and whose precocious celebrity, however, leaves only political traces. The young Évariste Galois, condemned a year ago because of a pronouncement at the Vendages des Bourgogne, fought against one of his old friends, a man as young as him, like him a member of the Society of Friends of the People, who had, in a recent interaction with him, figured equally in a political trial. It is said that love was the cause of the fight. The pistol was the weapon chosen by the two adversaries, but because of their old friendship they found it too difficult to have to look the one upon the other, and so they left the decision to blind fate. At close range, each of them was armed with a pistol and fired. Only one of the weapons was loaded. Galois was pierced through and through by the bullet

of his adversary; he was transported to the hospital Cochin, where he died after two hours. He was 22 years old. L. D., his adversary, is even a bit younger.[36]

The report was mistaken about the age of Galois. His funeral was attended by more than two thousand republicans. So what happened? I will not propose any conjectures. But I might as well note that Rothman himself, though rightly critical of arbitrary claims, proceeds to sketch his own conjecture: "With Stephanie's letters and the newspaper article we arrive at a very consistent and believable picture of two old friends falling in love with the same girl and deciding the outcome by a gruesome version of Russian roulette. This is my fairy tale. It has the virtues of simplicity and psychological truth. By comparison the tales of Bell, Hoyle and Infeld are baroque, if not byzantine, inventions."[37] At least Rothman clearly recognizes the gap between evidence and guesswork.

Invention is also the main cause of the legend of "the golden ratio," the irrational number 1.618 . . . , also known as "the divine proportion." Consider a line of a certain length L, say $L = 3$. If we cut it such that one part of it, p, is twice as big as the other, a (that is, $p = 2a$), then we have

$$\frac{L}{p} = 1.5, \text{ and } \frac{p}{a} = 2$$

The ratio of L to the bigger segment is larger than the ratio of p to a. However, in antiquity geometers found that there is a way to cut the line such that the ratio of the whole line L to the bigger segment b is *equal* to the ratio of the bigger to the smaller: $L/b = b/s$, or we can equally write

$$\frac{s + b}{b} = \frac{b}{s}$$

We might denote the value of these ratios by R. In the *Elements*, this geometric magnitude is called "the extreme and mean ratio," and it can be calculated in terms of three operations:

$$R = \frac{1 + \sqrt{5}}{2}$$

Numerically, the ratio is irrational, and its value is approximately

$$R = 1.61803398874 \ldots$$

Now, considering our initial line L, exactly where do we have to cut it, numerically, so that the ratio of its parts is R? If the length of $L = 1$ meter, then we should cut the line at 61.803 . . . centimeters of its length, such that it consists of two parts, one measuring .61803. . . m and the other .38196. . . m.

Thus we have

$$\frac{L}{b} = \frac{1}{0.61803\ldots} = \frac{b}{s} = \frac{0.61803\ldots}{0.38196\ldots} = 1.61803\ldots$$

These numbers have fascinating mathematical properties. For example, we have

$$1.61803\ldots \times 0.61803 = 1$$

Also

$$(1.61803\ldots)^2 = 2.61803\ldots$$

And also

$$(0.61803\ldots)^2 = 0.38196\ldots$$

Next, if we consider the sums $1 + 1 = 2$, $1 + 2 = 3$, $2 + 3 = 5$, $3 + 5 = 8$, and so on, then the ratios of successive pairs of numbers in this so-called Fibonacci series (known for many centuries before Fibonacci lived), 1, 1, 2, 3, 5, 8, 13, 21, 34, 55, . . . , converge toward

$$3/5 = .6,$$
$$5/8 = .625,$$
$$8/13 = .6153\ldots$$
$$13/21 = .6190\ldots$$
$$21/34 = .6176\ldots$$
$$34/55 = .6181\ldots$$
and so on.

This ratio is now known by the letter Greek letter φ (phi). Among its various other interesting properties, we have $\varphi^2 = \varphi + 1$ and also

$$\varphi = \cfrac{1}{1 + \cfrac{1}{1 + \cfrac{1}{1 + \ldots}}}$$

In 1509, a Franciscan friar and mathematician, Luca Pacioli, published a book about this interesting ratio. He called it "the divine proportion" because, he said, it shared five properties of God. Like God, this number was "only one and not more." Its alleged second property was that it "corresponds to the Holy Trinity. That is, just as in divinity there is one substance and three persons, Father, Son and Holy Ghost," the divine proportion involves three parts. The third commonality was that as "God cannot be defined or understood by us with words, in the same way our proportion can never be determined by an intelligible number nor expressed by any rational quantity, but is always hidden and secret, and mathematicians call it irrational." Next,

Pacioli claimed that just as God never changes and is in everything everywhere, so too the divine proportion can never change and is the same in all quantities large or small. The fifth correspondence was that just as God had created heavenly matter from "quintessence," the dodecahedron according to Plato, so too, Pacioli claimed, this figure of twelve pentagons has to be constructed using the divine proportion.[38]

Other mathematicians too became increasingly impressed by this ratio. In 1619, the mystic astronomer Johannes Kepler noted: "Today both the section, and the proportion it defines are given the title 'Divine,' because of the marvelous nature of the section and its multiplicity of interesting properties."[39] To Kepler, foremost among such properties was that once a line is divided by the divine section, if the larger segment is next added to the whole line, then the resulting compound line is divided again in the same proportion. In 1621, Kepler commented on its significance: "two of the treasures of Geometry: one, the relation of the hypotenuse in the right angle to the sides; and the other, the division of a line into extremes & mean ratio." And to this he added an author's endnote, saying (writers usually misquote this): "Two Theorems of infinite usefulness, precious value, but there is a great difference between the two. The former, that the sides of a right angle bear as much as the hypotenuse, I say that it resembles a mass of gold; the other, of the proportional section, you may call a Gem. It is beautiful in itself, but is worth nothing for the ends of the former, which promotes further knowledge."[40] Thus Kepler appreciated the divine ratio, while he valued the so-called theorem of Pythagoras as gold. Yet finally gold and the proportional section appeared in the same sentence.

For decades, many writers have claimed that the expression "golden section" first appeared in German, in a book of 1835 by Martin Ohm. But actually, that's a mistake; it's just the common fallacy of assuming that the earliest source one knows is its earliest origin. I don't know who first used that expression or when. But an earlier instance in Latin, "Sectio divina," appears in a book on mathematics from 1772, by a German Jesuit, Ignaz Pickel.[41] Previously, the expression "golden section" had been used for years, referring to the gilded edges of books and to a surgical operation.[42] By 1820, a German periodical on music, published in Vienna, alluded to "the well-known *Sectio divina geometrica* (also as *Sectio divina* or *Sectio aurea* . . .)."[43] In 1828, the expression "golden section" appeared in German in a schoolbook by a Dr. Ephraim Salomon Unger.[44]

Later, interest in the "golden section" grew greatly, thanks to the works of Adolf Zeising, a German psychologist who was interested in beauty. Zeising claimed that, among the ancient Greeks, Pythagoras was the first to recognize numbers as the unique expression of natural harmony and that Pythagoras saw in numbers "the epitome of perfection, the foundation of all virtue, and so also the source of all beauty."[45] Just as Pythagoras had apparently discovered the mathematical basis of beauty in musical harmonies, Zeising tried to find the mathematical basis of beauty in natural organisms.

In 1854, Zeising published his *New Theory of the Proportions of the Human Body*. He argued that the "golden section" was the universal law in the formation and structure of natural forms and of the most beautiful works of art. He called this "the aesthetic law of proportion" and argued that it appears in the arrangement of branches and stems in plants and the veins of leaves. He included illustrations exhibiting correspondence of proportions between parts of plants, leaves, and flowers.

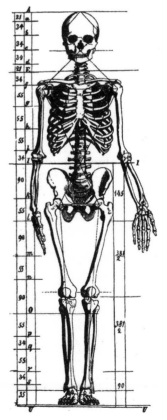

Zeising further argued that the golden section was beautifully manifest in the human body. Looking at the structure of the human skeleton, Zeising remarked that the gap beneath the lowest ribs and above the crest of the hip bones marks a clear partition of the human body; that the golden section falls within this gap; and that moreover, when the flesh is considered, the golden section corresponds to the navel.

Zeising remarked that the matching location of the golden section and the navel has an extraordinary significance. He wrote that "the inner nucleus and germ of the whole human being, and the navel, which is really the starting point of his existence, is a birthmark of his relation to the general." The upper part of the human body, he claimed, exhibits unity, while the lower part prominently exhibits separation, duality. The whole thus embodies a union of unity and duality and therefore "an image of the Trinity or an image of the

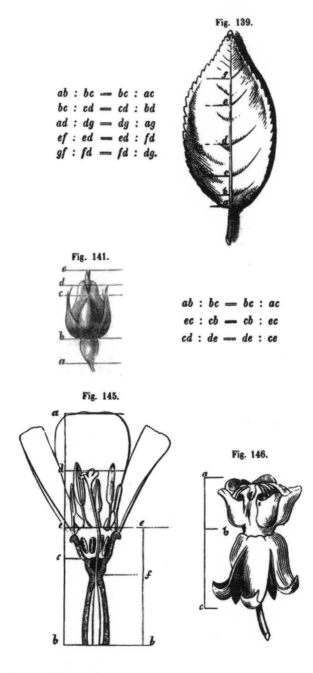

$$ab : bc = bc : ac$$
$$bc : cd = cd : bd$$
$$ad : dg = dg : ag$$
$$ef : ed = ed : fd$$
$$gf : fd = fd : dg.$$

$$ab : bc = bc : ac$$
$$ec : cb = cb : ec$$
$$cd : de = de : ce$$

Fig. 139.

Fig. 141.

Fig. 145.

Fig. 146.

Figure 5.2. Zeising claimed, for example, that the flower of the Asclepias Syriaca (fig. 146) shows that its petals and drooping sepals are divided by the golden section at *b*.

highest perfection or divinity." He thought that here was evidence of humans' likeness to God.[46]

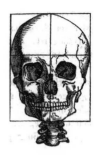

Zeising also analyzed the shape of the human skull in terms of divisions made to match the golden section. Enclosing first the skull in a rectangle, if the vertical sides of this rectangle are cut by a line placed at the golden section, the shorter, top part of the skull forms a half circle, he said. It seemed that Zeising's work had shown that there is a nonobvious mathematical order in the great varieties of natural forms, that there is a recurring unity of proportions between an organic whole and its parts.

Zeising's work was well received. Promptly, the expression "golden section" splashed into the English language. Irrespective of the truth of Zeising's claims, readers enjoyed finding the apparent hand of God in the mathematical structure of things. To give an early example, consider a British book review of Zeising's work, from 1855:

> Evidence, real or apparent, is of course discovered and set in order by the writer, whose enthusiasm is always in a glow, and whose hands and head never grow weary. The truth of the theory we leave to the arbitrement of those who are wise in aesthetics. One result we gratefully acknowledge. Our speculator has shown with fresh evidence and new illustrations that God's universe is full of beauty, and that the beauty of God's universe is full of design. He who has entered into the conception and recognised the force of this book can no longer think that beauty is an accident, but must own that it is an idea in the Platonic sense of the term, a preconception and its development. In other words the beauty of the universe is both an utterance and a reflex of the mind of its Author. The world is beautiful because it was made by God, the fountain of beauty, inasmuch as he is the fountain of symmetry and order. And the beauty of the world leads the mind to God, as to its Author, by links of thought which to the firmness of adamant add the brilliancy of gold.[47]

In 1509, Luca Pacioli had indirectly argued that architects should design some parts of their buildings in accord with the divine proportion.[48] Now Zeising further argued that the aesthetic law of proportion was manifest in some of the most beautiful works of art and in "not a few of the works of architecture from all periods, namely those that are recognized from time immemorial as masterpieces and models of beauty." He wrote that artists sometimes unconsciously design their works in ways that approximately match the golden section with surprising accuracy. He gave a few examples from ancient Greek architecture, starting with the famous Parthenon at Athens. Zeising claimed that the top of its columns is located at the golden section of the height of the Parthenon, from the bottom of its stairs to the top of its gable. He also argued that the dimensions of some cathedrals exhibit proportions similar to those given by the golden section.[49] It would be wrong to say that Zeising "discovered" the golden section in ancient architecture and art in general; it would be more accurate to say that he introduced or invented it. He proposed an attractive unity in the structure of nature, human bodies, and the arts. It served to redirect attention to the old notion of man as the measure of all things.

In the 1860s, Gustav Fechner began to experimentally test the effect of the golden section on people.[50] Fechner presented subjects with ten four-sided figures, cards, and asked them which looked most pleasant. The quadrilaterals varied from a square to a rectangle whose sides had a ratio of 2 to 5, that is, figures with aspect ratios from 1 to .4. The three rectangles in the middle, those with aspect ratios of .57, .62, and .67, were chosen by 76 percent of the subjects.[51] So it seemed that the rectangle with sides in the ratio $2.1 : 3.4 \approx 0.618 \ldots$ indeed was visually pleasant. Fechner doubted the general validity of Zeising's claims but still noted that the matter might involve a real discovery in aesthetics. Fechner tried to confirm people's apparent preference, so he later used ellipses with golden proportions, but he found no confirmation, so he cast more doubt on Zeising's claims and later even ridiculed them.[52]

Meanwhile, Pacioli's old suggestion that architects should mind the divine ratio, along with Zeising's recent but fictitious claim that architects in antiquity had actually used it, began to influence some actual working architects. One commentator soon complained that architecture was being infected, because "the use of the golden section . . . has apparently burst out into a sudden and devastating disease which shows no sign of stopping."[53] Claims about the golden section in architecture quickly multiplied. For example, supposedly this famous number was in the Great Pyramid of Egypt (ca. 2560 BCE), dividing its slant height by half of its base length.[54] Such claims were falsely attributed to Herodotus, who wrote no such thing.[55] And such claims were also architecturally false.[56] In 1978, one writer explained that it is problematic to attempt to reconstruct an architect's or an artist's theory of proportion, if they even had one, on the basis of measurements alone.[57] Regardless, some writers still falsely claimed, for example, that "the Egyptians also used the golden ratio, which the Rhind Papyrus refers to as a 'sacred ratio,' in building the Great Pyramid at Giza."[58] Thus Pacioli's "divine" ratio of 1509 became projected *forty-five hundred years* into the past!

In any case, the fascination with the golden ratio reached a most powerful way of influencing people: it began to be taught to children. In 1959, Walt Disney released a short animated cartoon film titled "Donald in Mathmagic Land."[59] The film was played on television, it gained wide distribution in schools, and it became one of the most popular educational films produced by Walt Disney. It is now freely available for instant viewing on the Internet, and many viewers have written comments about how much they loved it when they were young.

In the film, Donald Duck appears as an explorer, carrying a rifle as he enters a dark realm, a mysterious and haunting world of numbers. Nearly everything is made of numbers: a waterfall flows with numbers, and trees have square roots. Donald dreads mathematics as a subject for "eggheads," and so he tries to leave, but the narrator abruptly transports him "to ancient Greece, to the time of Pythagoras, the master egghead of them all, the father of mathematics and music." The narrator says that Pythagoras discovered the mathematical principles of harmony. Donald Duck then sees Pythagoras himself, a jolly, fat, bearded man playing a lyre along with other Pythagoreans. Donald Duck shakes hands with Pythagoras, who then vanishes. Right then, Donald sees that on the palm of his hand there has appeared the image of a black pentagram, which then becomes golden. At that point, the narrator says: "It was our old friend Pythagoras who discovered that the pentagram was full of mathmagic." He then illustrates how the ratios of lines in the pentagram produce "the golden section." He continues: "But this is only the beginning. Hidden within the pentagram is a secret for creating a golden rectangle, which the Greeks admired for its beautiful proportions and magic qualities." He soon adds: "To the Greeks, the golden rectangle represented a mathematical law of beauty. We find it in their classical architecture. The Parthenon, perhaps one of the most famous of early Greek buildings, contains many golden rectangles. The same golden proportions are also found in their sculpture. In the centuries that followed, the golden rectangle dominated the idea of beauty and architecture throughout the Western world. The cathedral of Notre Dame is an outstanding example. The Renaissance painters knew this secret well." All of these seemingly historical claims are false, yet this Disney film shows animated sequences in which golden rectangles are superimposed on illustrations of such famous buildings and human forms. Then abruptly, Donald Duck interrupts the sequence and tries to forcefully fit *himself* into golden rectangles. He struggles but fails. The narrator continues to show how pentagons, pentagrams, and the golden section (allegedly) show up in many natural shapes: flowers, starfish, spiral seashells, tree branches, pinecones, honeycombs, flies' eyes. The narrator comments: "The profusion of mathematical forms brings to mind the words of Pythagoras: 'Everything is arranged according to number and mathematical shape.' Yes, there is mathematics in music, in art, in just about everything—and as the Greeks had guessed, the rules are always the same."

Like Zeising, Disney productions claimed that the golden ratio was in the dimensions of the Parthenon.[60] But George Markowsky and others later found that such claims were arbitrary.[61] Writers further claimed that artists such as Leonardo da Vinci, Mondrian, and Seurat had used the golden ratio in their paintings.[62] Again, such claims lack substance.[63] Following Fechner's early work, some writers still think that people tend to aesthetically prefer rectangles that exhibit the golden ratio between their sides.[64] Yet numerous later systematic experiments have failed to confirm Fechner's apparent findings; they've shown, really, no preference for such "golden rectangles."[65] An online poll on the matter asked viewers to choose among rectangles, all having widths of 54 units but having various heights of 65, 68, 72, 77, 81, 87, 96, and 108 units. The sixth, $87/54 = 1.6111\ldots$, was close to a golden rectangle, but it was picked as the most pleasing rectangle by only 12 percent of 1,501 respondents. The 72 by 54 rectangle ($1.333\ldots$) won, garnering 30 percent of the votes. The percentages of respondents who chose each of the eight rectangles were, respectively, 9, 3, 30, 16, 19, 12, 5, and 8.[66]

Following Zeising, other writers claim to find the golden ratio in select dimensions of animals and human bodies, with scant evidence. One popular claim is that for most people their navel divides their height into the golden ratio.[67] Yet measurements of 319 individuals at Middlebury College in Vermont showed that most navels were higher.[68]

Commenting on this obsessive and delusional search for the golden ratio in art, architecture, and nature, the esteemed writer of mathematics Martin Gardner denounced it as a cult, a senseless devotion to an intellectual fad.[69] And he admitted that in the past he too had echoed such baseless claims. Regardless, some writers have claimed that Mozart used phi to divide his piano sonatas into parts. However, John Putz has convincingly showed that phi is not in Mozart's sonatas by giving a mathematical explanation of how misguided people could conclude that it was.[70]

Writers have also claimed that the golden ratio determines the spiral of various seashells, such as the chambered nautilus shell.[71] This has been used as a famous example of the beautiful connection between mathematics and nature. Hence, the image of a nautilus shell served for years as a popular illustration on the cover of mathematics books. But in 1999, a retired mathematician, Clement Falbo, took the time to actually *measure* nautilus shells in the collection of the California Academy of Sciences in San Francisco. The ratios of their widths ranged from 1.24 to 1.43, with an average of 1.33; that is, none was close to 1.618.[72] Despite his commendable and lucid analysis, however, Falbo still fell into a Pythagorean trap: he claimed that "as early as 540 BC the Pythagoreans had studied it [the golden ratio] in their work with the pentagon"—not true: there is no such evidence for that at all. And how did that claim arise? Through speculative inventions and associations.

The process of fictionalization grows in historical works, but it also gains impetus from literature. Stories about the golden ratio, for example, appear in the international bestselling novel *The Da Vinci Code*, by Dan Brown, first published in 2003. In it, the author echoes the old claim that the golden ratio, "generally considered the most beautiful number in the universe," is found in the proportions of nautilus shells, in the ratio of human height to navel height, in the Parthenon, in the Egyptian pyramids, and in various other places.[73] Brown adds that the proportion between female bees and male honeybees in any hive is always the divine proportion. False again: not always, not generally, not in any case.

The number serves as an alleged but satisfying way to interconnect nature, mathematics, art, and religion. Dan Brown writes: "The truly mind-boggling aspect of PHI was its role as a fundamental building block in nature. Plants, animals, and even human beings all possessed dimensional properties that adhered with eerie exactitude to the ratio of PHI to 1." And he echoes or invents further historical fictions to win his audience: "Nobody understood better than Da Vinci the divine structure of the human body. Da Vinci actually *exhumed* corpses to measure the exact proportions of human bone structure. He was the first to show that the human body is literally made of building blocks whose proportional ratios *always* equal PHI. . . . When the ancients discovered PHI, they were certain that they stumbled across God's building block for the world, and they worshipped Nature because of that."[74]

Brown's historical fictions are very instructive in how clearly they bring out the human urge to draw connections, to disguise historical ignorance with apparent certainties. We might well point out which statements are fictions and why, but it would be difficult to reach as many people as are influenced by the fiction. As of 2009, *The Da Vinci Code* had sold eighty million copies worldwide, having been translated into forty-four languages.

Moreover, the book gains credibility because it is often portrayed as historical. For instance, a news interviewer asked the author how much of the novel is true, and Brown replied: "99 percent of it is true. All of the architecture, the art, the secret rituals, the history, all of that is true, the Gnostic gospels. All of that is—all that is fiction, of course, is that there's a Harvard symbologist named Robert Langdon, and all of his action is fictionalized. But the background is all true."[75] This seems to be part of what makes the novel appealing to so many readers, the sense that they are learning about many real and interesting things. But really, there's a neglected process of invention at play here. And writers themselves seem unaware of it.

Just as Luca Pacioli noted that a property of God is to be everywhere, so too writers now claim to find phi in many places. They imagine and use the golden ratio in Web design, puzzles, photography, urban design, financial markets, poetry, and home building.[76] Even writers who are critically aware of the cult of the golden ratio still sometimes do not resist stating false claims about it. The cover art of one book, for example, shows a reproduction of Leonardo da Vinci's Mona Lisa with three golden-looking rectangles superimposed on it. One rectangle frames Mona Lisa's face, as if to illustrate Leonardo's use of phi. But it doesn't quite fit. Its right-hand side only almost goes to the start of her hair, and its lower side lies somewhat below her chin.[77]

As we have seen, the process of speculative invention is a main causal factor in the growth of other mythical stories in math. For example, it amazes me to see that some books on the history of math give a detailed geometrical analysis on the reputed pentagram of the Pythagoreans, for example, regarding how they "could have" constructed it and how they "might have" used it to derive incommensurability, or the golden ratio, or the notion of infinity—all fashioned upon a couple of brief old claims that say nothing of the sort. In the bestselling book *The Golden Ratio,* Mario Livio echoes the claim that it is "very plausible" that "perhaps" Hippasus of Metapontum discovered the golden ratio.[78] Such writers do not emphasize that the earliest mention (by Lucian) of the alleged pentagram *of health* of the Pythagoreans dates from *almost seven centuries* after the death of Pythagoras. Yet they arbitrarily ascribe it to Pythagoras's lifetime or to Hippasus. One free association easily leads to another, and guesswork becomes sold as history. Thus developed "Nature's Greatest Secret," the golden ratio.[79]

Eventually, the credit is allocated to Pythagoras himself. A religious attitude toward mathematics thus serves to convert an ancient religious leader into a mathematical genius. Thus, Mario Livio writes, "Pythagoras . . . spent endless hours over this simple ratio and its properties."[80] Finally, another bestselling book, *Zero: The Biography of a Dangerous Idea,* by Charles Seife, claims that "for Pythagoras, the golden ratio was the king of numbers."[81] Again, this is really just fiction, like the wonderful Disney cartoons.

In the popular stories about Gauss, Galois, and the golden ratio, we see how mathematicians, scientists, and writers tend to invent the past. These are hidden inventions, not anything that the storytellers portray openly as fiction. Likewise, there is another more important kind of invention that most mathematicians do not teach: invention in mathematics itself.

6

FROM NOTHING TO INFINITY

Mathematicians have the distinction of agreeing about results more often than members of most other professions. I think that mathematicians usually agree with one another more than the members of any science, any political party, and even any religion. But nevertheless, we can consider instances in which mathematicians have disagreed about various things, even the result of an operation. We would expect that performing basic operations on basic numbers such as 0, 1, 2, 3, would not yield disagreements. And it is well known that you cannot divide a number by zero. Math teachers write, for example, 24 ÷ 0 = undefined. They use analogies to convince students that it is impossible and meaningless, that "you cannot divide something by nothing." Yet we also learn, however, that we can multiply by zero, add zero, subtract it, and so on, and some teachers explain that zero is not really nothing, that it is just a number with definite and distinct properties. So why not divide by zero? In the past, many mathematicians did.

In ancient Greece, Aristotle argued that there exists no ratio of zero to a number.[1] But in 628 CE, the Indian mathematician and astronomer Brahmagupta claimed that "zero divided by a zero is zero."[2] At around 850 CE, another Indian mathematician, Mahavira, explicitly argued that any number divided by zero leaves that number unchanged, so then, 24 ÷ 0 = 24. Later, around 1150, the mathematician Bhaskara gave yet another result for such operations, arguing that a quantity divided by zero becomes an infinite quantity. This idea persisted for centuries; for example, in 1656, the English mathematician John Wallis likewise argued that $24 \div 0 = \infty$, using this curvy symbol for infinity, as he had done previously.[3] For $24 \div n$, for ever smaller values of n, the quotient becomes increasingly larger. For example, we have

$$\frac{24}{3} = 8, \ \frac{24}{2} = 12, \ \frac{24}{1} = 24, \ \frac{24}{.5} = 48, \ \frac{24}{.1} = 240, \ \frac{24}{.0001} = 240,000$$

and so forth. Therefore, Wallis argued that the quotient becomes infinity when we divide by zero.

The common attitude toward such old notions is that past mathematicians were plainly wrong, confused, or "struggling" with division by zero. But that attitude disregards the extent to which even formidably skilled mathematicians thoughtfully held such notions. In particular, the Swiss mathematician Leonhard Euler is widely admired as one of the greatest mathematicians in history, having made extraordinary contributions to many branches of mathematics, physics, and astronomy in hundreds of masterful papers written even during years of blindness. Euler's *Complete Introduction to Algebra* was published in 1770, in German (a Russian translation had appeared previously). This textbook has been praised as the most widely published book in the history of algebra. We here include a page from an English translation, in which Euler discusses division by zero. He too argues that it gives infinity.[4]

A common and reasonable view is that despite his fame, Euler was clearly wrong, because if any number divided by zero is infinity, then all numbers are equal, which is ridiculous. For example:

if $\quad 3 \div 0 = \infty, \quad$ and $\quad 4 \div 0 = \infty,$

then $\quad \infty \times 0 = 3, \quad$ and $\quad \infty \times 0 = 4$

Here a single operation, $\infty \times 0$, has multiple solutions, so apparently, $3 = 4$. Therefore, one might imagine that there was something "premodern" in Euler's *Algebra,* that the history of mathematics includes long periods in which mathematicians did not find the right answer to certain problems.

In 1828, the German mathematician Martin Ohm discussed division by zero as follows: "If a is not zero, yet b is zero, then the quotient $a : b$ or a/b has no meaning, because any difference between whole numbers $\mu - v$, multiplied with b, i.e., with zero, only gives zero, therefore it cannot give a, so long as a is not zero." Thus he argued that division by zero is prohibited, and he noted that "0/0 has infinitely many meanings, i.e., a fully undetermined meaning."[5] Ohm emphatically instructed his readers: "*Never divide by zero.*"[6] Subsequently, other writers also increasingly argued that division by zero is meaningless. Writers also increasingly defined division in terms of multipli-

more particularly as it will be of the greatest importance in the following part of this treatise.

We may here deduce from it a few consequences that are extremely curious and worthy of attention. The fraction $\frac{1}{\infty}$ represents the quotient resulting from the division of the dividend 1 by the divisor ∞. Now we know, that if we divide the dividend 1 by the quotient $\frac{1}{\infty}$, which is equal to nothing, we obtain again the divisor ∞ : hence we acquire a new idea of infinity; and learn that it arises from the division of 1 by 0; so that we are thence entitled to say, that 1 divided by 0 expresses a number infinitely great, or ∞.

84. It may be necessary also in this place to correct the mistake of those who assert that a number infinitely great is not susceptible of increase. This opinion is inconsistent with the just principles which we have laid down; for $\frac{1}{0}$ signifying a number infinitely great, and $\frac{2}{0}$ being incontestably the double of $\frac{1}{0}$, it is evident that a number, though infinitely great, may still become twice, thrice, or any number of times greater*.

---

* There are other properties of *nothing* and *infinity* which it may be proper to notice in this place.

1. Nothing, added to or subtracted from any quantity, makes it neither greater nor less.

2. Any quantity multiplied by 0, that is, a quantity taken no times, gives 0 for a product; or $a \times 0 = 0$.

3. $a^0 = 1$, whatever be the numeral value of a. For $a^0 \times a$ $a^{0+1} = a^1 = a$; but $1 \times a = a$ likewise; therefore $a^0 = 1$.

4. Since $\frac{a}{0} = \infty$, therefore $0 \times \infty = a$; that is, nothing multiplied by infinity produces a finite quantity.

Figure 6.1. An English edition (1810) of Leonhard Euler's *Algebra*, showing why Euler argued that division by zero is possible. Although this translation was based on a French edition, rather than on the German original (1770), it is very accurate. The footnotes, however, are not in the original; they were added by a later editor.

cation, such that "the quotient" of a/b is a number x such that $a = bx$, thus excluding division by zero by definition.[7]

The number zero is extremely interesting precisely because of its uniqueness. In the words of Constance Reid, "zero is the only number which can be divided by every number, and the only number which can divide no other number."[8]

Table 6 summarizes the various views on division that we've mentioned.[9] A glance at this kind of list might give the impression that there transpired a prolonged time of confusion prior to the juncture when mathematicians finally recognized the correct solution: that the result of division by zero is *not* infinity but is undefined. Historically, the expression "undefined" is very appropriate because it is not that the result of the operation has never been defined—actually it used to be explicitly defined—but that later mathematicians chose to "undefine" it. And looking at the list, someone might want to draw a sharp distinction somewhere between Euler and Ohm. But I disagree with this interpretation that seems to divide the past into two periods, the premodern and the modern, because the results of ambiguous operations might not be as settled as they appear.

Table 6. Division by Zero, Selected Answers across Several Centuries: Old Confusions or Substantial Disagreements?

ca. 340 BCE	Aristotle	1 : 0 does not exist
ca. 628	Brahmagupta	$1 \div 0 = \frac{1}{0}, 0 \div 0 = 0$
ca. 830	Mahavira	$1 \div 0 = 1$
ca. 1150	Bhaskara	$1 \div 0 =$ infinity
1656	John Wallis	$1 \div 0 =$ infinity
1770	Leonhard Euler	$1 \div 0 =$ infinity
1812	Sylvestre Lacroix	$1 \div 0 =$ infinity
1828	Martin Ohm	$1 \div 0$ is meaningless
1830	George Peacock	$1 \div 0 =$ infinity
1831	Augustus De Morgan	$1 \div 0 =$ infinity
1881	Axel Harnack	$1 \div 0$ is impossible
1896	Louis Couturat	$1 \div 0$ is impossible
1896	Charles Smith	$1 \div 0$ is meaningless, impossible, *or* infinity
1911	E. B. Wilson	$1 \div 0$ is impossible
1928	Konrad Knopp	$1 \div 0$ is undefined, impossible, meaningless

Euler had fair reasons for his arguments. If division by zero gives infinity, we note, then $\infty \times 0$ has multiple solutions, which seems to be an unforgivable problem. Yet the notion of multiple solutions to one equation does not seem impossible. Euler argued, for one, that the operation of extracting roots yields multiple results.[10] For example, the cube root of 1 has three values, two of which are imaginary.

$$\sqrt[3]{1} = 1, \quad \sqrt[3]{1} = \frac{-1 + \sqrt{-3}}{2}, \quad \sqrt[3]{1} = \frac{-1 - \sqrt{-3}}{2}$$

Even today, all mathematicians agree that root extraction yields multiple results. They just do not equate the resulting values. So why not also admit multiple results when multiplying zero by infinity?

An alternative way to understand the historical disagreements over division by zero is to recognize that there is change in mathematics, that certain mathematical operations evolve over time. In antiquity, mathematicians did not divide by zero. Later, some mathematicians divided by zero, obtaining either zero or the dividend, for example, that $24 \div 0 = 24$. Next, other mathematicians argued, for centuries, that the correct quotient is actually infinity. And nowadays again, mathematicians teach that division by zero is impossible, that it is "undefined." But ever since the mid-1800s, thanks to works by George Peacock and others, algebraists realized that certain aspects of mathematics are established by convention, by definitions that are established at will and are occasionally refined or redefined. If so, might the results of division by zero change yet again? As we are taught in school that division by zero is undefined, it might seem impressive that mathematicians have maintained this for roughly *a century.* However, how is that impression affected by knowing that previously, for roughly *eight hundred years,* many mathematicians argued that the result is infinity? Might the result change yet again?

To answer this question, we need to consider not only the concept of zero but the operation of division itself. This operation has been defined in various ways throughout history. What do we mean when we say that we will divide something? Intuitive notions include ideas about "cutting up," "distributing," "fitting into," and so on. Are these all physical operations? When we divide, are we just representing physical processes? Or isn't mathematics supposed to deal essentially with abstract entities and relations?

Many basic operations are rooted in physical experience and procedures, but as they became refined as abstract mathematical concepts, they were extended in ways that are not necessarily physical or not exclusively physical. Thus, for instance, it became algebraically plausible to divide by zero, even if physically the operation seemed to lack meaning. After all, algebraists, such as Girolamo Cardano in the 1540s, sought to develop a universal algebra, an algebra that would provide an answer to any problem that could be formulated in its symbols.[11] Accordingly, if we can divide 24 by 2, why not by 0?

A similar issue arose in division with negative numbers. First we might conceive of negative numbers as representations of experience, for example, as debts. But once such numbers were admitted into mathematics, they came to occupy positions where they did not seem to have evident meaning. For example, if −5 is a debt of five dollars, then what is −5 ÷ −4? How do you divide a debt of five dollars by a debt of four dollars? The question seems meaningless, yet the drive to develop a universal algebra requires that we provide answers to problems such as −5 ÷ −4, even if we have to leave some practical ideas behind, such as the idea that a negative always represents a debt.

Likewise, division used to mean, "How many times does one quantity fit in another?" And it still does, early in grade school. But once negative numbers became admitted as dividends and divisors, that meaning too became problematic. For example, consider the expression: 16/4 = ? We might read this as, "How many times does the quantity four fit into sixteen?" But consider now:

$$\frac{aa}{-a} = -a$$

Can we again ask, how many times does −a fit into aa? Accordingly, in a memoir of 1780, the French polymath Jean d'Alembert argued: "If someone asks why aa/-a = -a, I will reply that by dividing the quotient of the division of aa by −a, one does not ask how many times −a is contained in aa, that which would be absurd, one asks a quantity such that being multiplied by −a gives aa."[12]

Here, division is redefined not to consist of "fitting into" but instead as a sort of inverse of multiplication. But wait—is division really the inverse of multiplication? Again, the problem is that if 5 × 0 = 0, then we would expect that the inverse operation would allow us to do 0 ÷ 0 = 5. But if division by zero is impossible, then division is not really the inverse of multiplication: it does not undo an infinity of possible multiplications.

Thus mathematicians, over the centuries, revised the definitions of some basic operations, modifying them especially in relation to new elements, such as strange numbers. D'Alembert discussed another apparent paradox of division. We are all comfortable with the proposition

$$\frac{1}{2} = \frac{3}{6}$$

Both pairs of quantities have the same ratio. We all also understand and accept the inequality

$$\frac{1}{2} \neq \frac{2}{1}$$

The ratio of the smaller number to the greater is not equal to the ratio of the greater to the smaller. But consider the following ordinary proposition of mathematics:

$$\frac{-1}{1} = \frac{1}{-1}$$

How can the ratio of a smaller quantity to a larger be the same as the ratio of the larger quantity to the smaller? Or, how can one possibly divide a smaller number by a greater and get the same result as by dividing the greater number by the smaller? What does this proposition mean physically, say, in terms of apples? D'Alembert concluded that this relation shows that negative numbers are not less than zero. But most mathematicians disagreed. (Still, we can understand d'Alembert's conclusion, for example, by interpreting negatives and positives as motions to the left or to the right, in which case, really, it makes no sense to regard one displacement as less than the other.) Mathematicians' solution was to exempt numbers and operations from any necessary meaning that we commonly ascribe to them in everyday life. For example, we now say that we are not actually "cutting up" 1 into −1 parts.

Furthermore, division involves certain peculiarities when it is considered in relation to other operations. Consider the addition of fractions. Suppose that in one game, a baseball player goes to bat three times and manages to hit the ball once. Thus, his batting average for that game is 1/3 = 0.333 . . . In another game, the same player goes to bat four times and hits the ball once. His average for that second game is 1/4 = 0.25. Now, a baseball fan might want to know the batting average of this baseball player. The rules of algebra teach us that to add fractions such as 1/3 and 1/4 we need to first find their common denominator:

$$\frac{1}{3} + \frac{1}{4} = \frac{4}{12} + \frac{3}{12} = \frac{7}{12} = 0.5833 \ldots$$

So it would seem that the player's batting average is almost 0.6. That is an amazing result, and it is also nonsense. All baseball fans know that to calculate the average what we do instead is to add the numerator and the denominator directly (or the dividend and the divisor) and then divide

$$\frac{a}{b} + \frac{c}{d} = \frac{a+c}{b+d}$$

By this procedure we obtain a batting average of .2857 instead. And even if we divide the former value, .5833, by 2, given that there were two games, we still do not obtain .2857, but instead .2916. So what is the correct way to calculate the average?

Everyone agrees that in baseball, fractions should not be added according to the ordinary rule for adding fractions. Instead, we add the numerators directly and divide that by the sum of the denominators. Now, suppose that we construct a system of numerical rules in which we add fractions in this unusual way. Mathematician Morris Kline explained that this sort of "baseball arithmetic" involves some rules that are identical to the usual rules of algebra, while it also involves others that are distinct.[13] Yet in the context of its particular field of application, baseball, it would produce useful and coherent results. Thus, one might define division in a way that differs from the usual.

Return now to division by zero. At the University of Texas at Austin, I often teach a course to students who are majoring in mathematics and sciences. I sometimes ask them why division by zero is undefined, and I ask them to consider whether infinity is a plausible answer instead. Many of them reply that the latter approach does not make sense, and they remain very skeptical of arguments by past mathematicians, such as Wallis and Euler. Most students think that people of the past were just confused and that the solution of a basic operation such as division by zero has been clearly and permanently settled, presently, that it cannot and will never change. Still, in 2005, after discussing past approaches to division by zero, we considered what answer a computer would give. So we had the computer in our classroom, an Apple iMac, carry out division. I typed 24 ÷ 0, paused to let them see the numbers projected on the wall, and then hit the enter key. The computer replied "Infinity!" Surprise. Strange that it had an exclamation mark, as if it were yelling.

Some students complained that the computer was an Apple instead of a Windows PC. The following year, a similar Apple computer, a newer model, also answered "infinity," but without the exclamation mark. In 2010, the same operation, on a newer computer in the classroom, replied: *"DIV BY ZERO."* Personally, I think that *"DIV BY ZERO"* is not an answer. I feel cheated. I *know* the operation that I have keyed: what I want is the answer. Yet that same Apple computer has an additional calculator, a so-called scientific calculator, and the same operation on this more sophisticated calculator gave "infinity." Students' calculators, such as on their cell phones, gave other results: "error" or "undefined." One student's calculator, a Droid cell phone, answered "infinity." None of these answers is an accident; each has been thoughtfully programmed into each calculator by mathematically skilled programmers and engineers, presumably people trained at top institutions such as MIT and Caltech. Some anonymous crews of savvy programmers consciously decided that "infinity" was the correct answer for this kind of operation.

Computer scientists confront a basic and old algebraic problem: using variables and arithmetical operations, occasionally computers encounter a division in which the divisor has a value of zero—what should it do then? Stop? Break down? One professor of computer science who presently argues that division by zero really should produce infinity is James Anderson, at the University of Reading, in Berkshire, England. In a 2006 BBC television interview, Anderson explained: "Imagine you're landing, on an airplane. The automatic pilot is working. If it divides by zero the computer stops working, you're in *big* trouble; the same with the engine management system in your car. If it divides by zero, your car's not going to start in the morning. Heart's pacemaker: divides by zero, you're dead."[14] Writing on a whiteboard, Anderson teaches students to divide by zero, and he also divides zero by zero. I cannot resist quoting the words of the television news reporter, as the young students walked into the classroom: "School's in, but no simple algebra here. This lot is solving a problem that troubled the likes of Newton and Pythagoras. How to divide a number by zero."[15] As usual, the image of Pythagoras creeps in to add a semblance of history, while truly, there is not a shred of evidence that Pythagoras of Samos ever tried to divide anything by zero.

Likewise, some recent writers have ascribed the roots of the mathematical notion of infinity to, of all people, Pythagoras, again without any evidence. Amir Aczel writes: "But the roots of infinity lie in the work done a century before Zeno by one of the most important mathematicians of antiquity, Pythagoras." Aczel speculates that the reputedly Pythagorean pentagram pertained to infinity: "The Pythagoreans had a symbol—a five-pointed star enclosed in a pentagon, inside of which was another pentagon, inside it another five-pointed star, and so on to infinity." To the contrary, ancient evidence does not show that any Pythagoreans inscribed pentagrams into pentagons successively. At around 180 CE, Lucian merely claimed that they interlaced three triangles to make a pentagram as a symbol of health. Regardless, Aczel further speculates: "The Pythagoreans considered one as the generator of all numbers. This assumption makes it clear that they had some understanding of the idea of infinity, since given any number—no matter how large—they could generate a larger number by simply adding one to it."[16] Once again, both for zero and for infinity, recent commentators have arbitrarily credited Pythagoras.

Anyhow, the BBC published an online article to accompany the local news telecast about James Anderson, and it allowed readers to post comments. Since then, many hundreds of people have posted their comments, reacting to Anderson's claims. Many readers have replied that division by zero is clearly "impossible." Some have complained that the everyday examples Anderson gives are defective or "obviously ridiculous" because airplanes and heart pacemakers have control mechanisms that are programmed to handle exceptions, to prevent internal calculators from dividing by zero.

Actually, there are instances in which division by zero really has caused technical problems. For example, on 21 September 1997, the USS *Yorktown* battleship was testing "Smart Ship" technologies off the coast of Cape Charles, Virginia. At one point, a crew member entered a set of data that mistakenly included a zero in one field, causing a Windows NT computer program to divide by zero. This generated an error that crashed the computer network, causing failure of the ship's propulsion system, paralyzing the cruiser for more than a day.[17]

Nowadays, most people are quite comfortable with the answer "undefined," as if *a word* were a proper answer, as if there are some operations that are forbidden, impossible. But can somebody still come along and choose to *define* a meaning for this operation? For centuries, plenty of individuals seized the liberty to do just that.

Aside from old examples such as Wallis, Euler, and Peacock, a few more recent writers have sometimes acknowledged that a quotient of infinity is not entirely nonsense. For example, in the highly praised book *What Is Mathematics?* the authors Richard Courant and Herbert Robbins argue: "Expressions like 1/0, 3/0, 0/0, etc. will be for us meaningless symbols. For if division by 0 were permitted, we could deduce from the equation $0 \times 1 = 0 \times 2$ the absurd consequence $1 = 2$. It is, however, sometimes useful to denote such expressions by the symbol ∞ (read, 'infinity'), *provided that one does not attempt to operate with the symbol ∞ as though it were subject to the ordinary rules of calculation with numbers.*"[18] Accordingly, in a good textbook published in 2005, mathematician John Stillwell explains that this operation is sometimes used in projective geometry. He writes: "You remember from high-school algebra that division by zero is not a valid operation, because it leads from true equations, such as $3 \times 0 = 2 \times 0$, to false ones, such as $3 = 2$. Nevertheless, *in carefully controlled situations,* it is permissible, and even enlightening, to divide by zero. One such situation is in projective mappings of the projective line." Stillwell explains that although the function $f(x) = 1/x$ does not map all points on the real line R (because it does not map the point $x = 0$), it does work properly on the real projective line R $\cup$ {∞}, a line together with a point at infinity, such that "the rules $1/∞ = 0$ and $1/0 = ∞$ simply reflect this fact."[19] Stillwell explains that division by zero is useful in the more general context of linear fractional transformations. Meanwhile, a very few recent textbooks also give infinity as the quotient of division by zero.[20] Some books on operations research, also known as management science, also divide by zero as if it gives infinity.[21] Various computer books define division by zero as infinity.[22] Some computer programs handle $1/0$ not by returning an exception error, but by giving a specific answer, not infinity, but the largest floating point value possible in the particular system.[23]

Table 7 shows the results given for division by zero by various sources, including especially the more recent minority view that it is infinity.[24] This table does not convey the degree to which the majority of people, after the 1920s, accepted that this operation is "undefined." Since then, for every example listed stating that the quotient is infinity, we might guess that there are thousands of others stating that it is undefined. Instead, the aim here is merely to show that the old notion, that division by zero is infinity, did not disappear after Martin Ohm.

By this point one might wonder: so what? What are the consequences of using one word or another, "error," "infinity," and so on, as the answer for divide by zero? In fact, a few authors use the symbol for infinity ∞ as synonymous with "undefined."[25] The difference is between forbidding an operation and therefore carrying out ways to avoid it versus trying to define a system in which the basic rules apply to all numbers, without exceptions. The latter is an interesting notion, but difficult to realize in practice.

Return now to the two calculators in an Apple computer. It is interesting to compare how these two calculators handle arithmetic. Following division by zero, one calculator gives "*DIV BY ZERO.*" Keeping that expression on the screen, I next type [+] and [2] and it replies: "*ERROR.*" By contrast, the scien-

Table 7. Division by Zero, Selected Answers across Several Centuries: Old Confusions or Substantial Disagreements?

ca. 1150	Bhaskara	$1 \div 0$ = infinity
1656	John Wallis	$1 \div 0$ = infinity
1770	Leonhard Euler	$1 \div 0$ = infinity
1828	Martin Ohm	$1 \div 0$ is meaningless
1830	George Peacock	$1 \div 0$ = infinity
1881	Axel Harnack	$1 \div 0$ is impossible
1902	Arnold Emch	$1 \div 0$ = infinity
1928	Konrad Knopp	$1 \div 0$ is undefined, impossible, meaningless
1941	Richard Courant	$1 \div 0$ = infinity (sometimes is useful)
1986	Hewlett-Packard calculator	$1 \div 0$ = error
2005	John Stillwell	$1 \div 0$ = infinity (sometimes is useful)
2005	Apple iMac computer	$1 \div 0$ = infinity!
2006	Motorola cell phone	$1 \div 0$ = error
2006	James Anderson	$1 \div 0$ = infinity
2010	Droid cell phone	$1 \div 0$ = infinity
2010	Apple iMac computer dashboard calculator	$1 \div 0$ = div by zero
2010	Apple iMac computer scientific calculator	$1 \div 0$ = infinity

tific calculator, following 1/0, gives "Infinity," and I next type [+] and [2], and it replies "Infinity" again. Also, if we first do $1 - 2$, this calculator gives –1; by typing [M+] to store this value, we next do 1/0 to get "Infinity," on which we can immediately operate by doing Infinity $\times$ –1 (using the MR key, memory recall), and the calculator replies: "–Infinity."

In 2003, Eli Maor, of Loyola University in Chicago, published an article titled "Thou Shalt Not Divide by Zero!" He echoed the usual reasons why 1/0 is undefined: that it entails 1 = 0; that 0 fits infinitely many times in 1 though ∞ is not a real number; and that the expression $y = 1/x$ tends to $+\infty$ or to $-\infty$ depending on whether $x \to 0$ through positive or negative values, so the results diverge. Still, Maor proposed a way to overcome these objections. Historically, mathematicians overcame their distrust of imaginary and complex numbers by giving them geometric interpretations. Accordingly, Maor proposed a geometric meaning for ∞ and division by zero.[26] Consider a number line, x, which includes all the real numbers as usual. Maor argued that we can define a "number circle" that includes all the real numbers as well. The

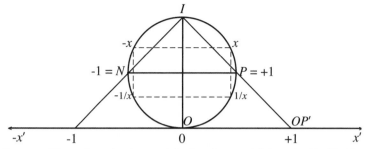

Figure 6.2. The real number line and the "number circle" described by Eli Maor.

point 0 on the number line coincides with the point zero in the number circle. Likewise, we can draw a straight line connecting any point on the number line to the point I on the circle, such that to any number P' on the line there corresponds only one point on the circle: the point of intersection, such as P.

For any other point on the number line we can similarly find a corresponding image point on the circumference, so that the circle includes every number on the number line, all ordered in the same sequence. Yet there is *one* point on the circle that does not correspond to any point on the number line, namely I. Maor suggested that if we set $P = +1$, and $N = -1$, then the operation of taking the additive inverse $-x$ of any real number x has the simple geometric meaning of reflecting any image point along the vertical diameter OI. Likewise, the operation of taking the multiplicative inverse $1/x$ reflects any image point on the circle in the horizontal diameter NP. Then, if we take the multiplicative inverse of 0, we get the point I, which Maor chose to designate by the infinity symbol. Therefore he wrote:

$$1/0 = \infty, \quad 1/\infty = 0$$

and he added: "Thus in one stroke it seems to have removed the taboo against zero division! We may again invoke an analogy with complex numbers: although these numbers had been in use since the seventeenth century, it was not until Argand, Wessel, and Gauss, around 1800, showed how operations with them could be interpreted geometrically, that mathematicians felt completely at ease with these once strange creations." Indeed, before imaginary numbers were interpreted as perpendicular to the real number line, mathematicians for centuries had ridiculed imaginary and complex numbers as: "false," "absurd," "nonexistent," "unintelligible," "impossible," "nonsense," "mistaken," and so on.[27]

Next, Maor postulated the following assumptions:

$x + \infty = \infty$, for all $x \neq -\infty$,

$x \times \infty = \infty$, for all $x \neq 0$,

$0 \times \infty = 1$

Thus infinity, like the number zero, would have distinct properties absent from other numbers. Maor then presented several parallel properties of zero and ∞, as portrayed in table 8.

Table 8. Algebraic Properties of Multiplication with Zero and Infinity, in the System of Rules Invented by Eli Maor

Equation	Solution Set	Equation	Solution Set
$0 \times x = 0$	all $x \neq \infty$	$\infty \times x = \infty$	all $x \neq 0$
$0 \times x = 1$	$x = \infty$	$\infty \times x = 1$	$x = 0$
$0 \times x = a$ $(a \neq 0$ or $1)$	$\varnothing$	$\infty \times x = a$ $(a \neq \infty$ or $1)$	$\varnothing$

He also specified opposites and reciprocals as follows:

$-0 = 0 \qquad -\infty = \infty$

$1/0 = \infty \qquad 1/\infty = 0$

But then Maor pointed to several difficulties in his proposed system. We might have expected that: $-\infty + \infty = 0$. But the usual notion that $\infty + x = \infty$, as stated above, leads to the following problem: if we let $x = -\infty$, we then have $-\infty + \infty = \infty$, instead of 0. Maor accepts both statements about the addition of positive and negative infinities, and therefore he reaches a contradiction, that $0 = \infty$.

At this juncture, however, Maor could have just said that in his system $-\infty + \infty \neq 0$, and he would then have avoided the contradiction. Notice also that Maor's definition of $-\infty = \infty$ differs from what was used in the scientific calculator above: $-\infty = -1 \times \infty$. Another alternative would be to reject, in Maor's system, the usual notion that $\infty + x = \infty$. After all, if we wish to explore a system in which $1/0 = \infty$, we should not be bound by all the typical rules that apply in the standard system, in which $1/0$ is impossible and undefined.

Regardless, Maor promptly stopped his inventive experimentation with symbols, because it clashed with notions that he expected the new system to have. Mathematics is an interesting and convincing mixture of propositions that seem clearly true and necessary and other propositions that seem counterintuitive and mysterious, but that cohere strongly with the former.

When mathematicians find that a new rule clashes with a prior and intuitive notion, they often lose interest in the new rule or concept; they think that it would come at the price of sacrificing large pieces of an established and reliable system.

Return now to computer scientist James Anderson, in England. Instead of arguing that infinity is a point outside the number line, he takes the real number line and adds two more numbers, $+\infty$ and $-\infty$, so that "the positive side goes all the way to infinity, and the negative side all the way to minus infinity." He calls this the "transreal" number line. Next, he posits these definitions:

$$\infty = \frac{1}{0}, \ -\infty = \frac{-1}{0}, \ \Phi = \frac{0}{0}$$

Anderson defines $0/0$ as "nullity," a new number that lies outside the transreal number line. By contrast, mathematicians have traditionally argued that $0/0$ is indeterminate, as explicitly argued long ago by Lacroix, Ohm, Peacock, De Morgan, and others. Yet by defining $0/0$ as a new number, Anderson claims to solve problems that otherwise lack definite solutions. For example, he handles zero to the power of zero as follows:

$$0^0 = 0^{1-1} = 0^1 \times 0^{-1} = \left(\frac{0}{1}\right)^1 \times \left(\frac{0}{1}\right)^{-1}$$

and since any number to the power of -1 is its inverse, he writes:

$$0^0 = \frac{0}{1} \times \frac{1}{0} = \frac{0 \times 1}{1 \times 0}$$

such that

$$0^0 = \Phi.$$

By contrast, many mathematicians have argued that 0^0 is equal to 1 or that it is indeterminate.[28] In response to the BBC's report of Anderson's "discovery," most viewers who posted comments were unimpressed and annoyed. Their criticisms include the following words, repeatedly: "ridiculous," "stupidity," "dumb," "nonsense," "impossible," "imaginary," "crazy," "absurd," "meaningless," "completely unnecessary," "insanity," "irrational," "totally worthless," "shame on the BBC." A recurring complaint is that Anderson just "invented" nullity; as one reader briefly put it: "Inventing new numbers, that's just cheating." Another recurring criticism is that "the idea of a point outside the number line is ridiculous." But then again, that was done with imaginary numbers, and mathematicians accepted it.

Despite the hundreds of bitter and colorful criticisms of Anderson, there seems to be no inconsistency in what he proposes. But still, his approach does not appear to give methods or solutions that mathematicians have yet considered useful. Previously, the expression 0^0 was not quite unsolved, but it was certainly ambiguous. Some mathematicians, especially those working with limits and continuous functions, leave 0^0 as undefined, indeterminate. Other mathematicians, some working in set theory and in contexts that do not require continuity, have instead defined $0^0 = 1$. Also, in the differential calculus, the power rule

$$\frac{d}{dx}x^n = nx^{n-1}$$

is not valid for $n = 1$ at $x = 0$, unless $0^0 = 1$. If instead we choose to adopt Anderson's nullity as the solution of 0^0, it is unclear whether that is somehow more useful than the previously defined values.

Anderson's proposed system is quite similar to the numerical system that is commonly used in most computers. In 1985, the Institute of Electrical and Electronics Engineers (IEEE) established a system of binary arithmetic known as the IEEE 754. It includes values of infinity, negative infinity, negative zero, and NaN (Not a Number), and it defines division by zero. The IEEE is a professional organization of engineers, incorporated in New York, that includes thirty-eight societies and seven technical councils and has roughly four hundred thousand members in more than 160 countries.[29] All numbers in the IEEE 754 system include a sign, positive or negative, so therefore this system includes a -0. Negative zero echoes a concept from mathematical analysis, namely that of approaching 0 from below as a one-sided limit. This number is also often used to represent negative values that are too small to be represented in a computer's limited precision. Also, computer scientists use it in

$$\frac{1}{-\infty} = -0, \text{ and } \frac{1}{-0} = -\infty$$

Meanwhile, the IEEE 754 system also involves so-called NaN data values, which serve to represent expressions such as $0/0$, 0^0, $\log -1$, and $\sqrt{-1}$. Since NaN can have various or indeterminate values, it is described as not being equal to itself.

Table 9 compares various approaches to division by zero and so forth. Some of Anderson's many critics have claimed that by introducing nullity Anderson simply gave another name to the well-known concept of Not a Number. In response, Anderson explained that the difference is that nullity *is* a number and that therefore it can be used consistently in mathematical

Table 9. Comparison of a Few Approaches to Zero and Infinity

	Standard mathematics	*Eli Maor's experiment*	*IEEE 754 binary arithmetic*	*James Anderson's transreal arithmetic*
-1×0	0	0	-0	0
$1/0$	undefined	∞	∞	∞
$-1/0$	undefined	∞	$-\infty$	$-\infty$
$0/0$	indeterminate	indeterminate	NaN	Φ
0^0	indeterminate or sometimes 1	unspecified	NaN	Φ
$\infty \pm 1$	∞	∞	∞	∞
$\infty \times 1$	∞	∞	∞	∞
$\infty \times -1$	often $-\infty$	∞	$-\infty$	$-\infty$
$\infty \times 0$	∞, or often 0 in probability theory	1	NaN	Φ
$\infty \times \infty$	∞	∞	NaN	∞
Other	$-\infty = \infty$ in some contexts	$0 = \infty$ contradiction	NaN $\neq$ NaN	$\Phi = \Phi$

operations. Likewise, in contemporary real analysis, ∞ is not a number, but a special limit that does not obey the axioms of real arithmetic, whereas in Anderson's arithmetic, ∞ and $-\infty$ are defined as numbers and are consistently distinct from one another, and $1/0$ and $-1/0$ are actual fractions.

In IEEE 754 computer arithmetic, an operation such as division by zero is still often described as an invalid operation, but nonetheless, it yields a definite and distinct binary result. Thus in the case of division by zero, we witness the recent evolution of an operation: decades ago it was considered impossible, and then it was called "undefined"; later that word was sometimes given other names, such as "infinity" and the symbol ∞, and nowadays such words and symbols are increasingly treated like numbers, by defining operations on them with distinct results, whether generally or at least in specific contexts. In one of his papers, Anderson comments on the history of numbers: "At each stage in the development of mathematics it is tempting to assume that one's own mathematics has achieved the pinnacle of success, but this is rarely, if ever, the case. . . . Human psychological limits have affected the structure of real arithmetic."[30] Operations that seem impossible continue to be avoided as impossible until some eccentric writers insist on defining a result to the operation, regardless. This happened with zero, irrationals, negative numbers, imaginaries, and so on, and of course Anderson wants nullity to be accepted as a legitimate number, eventually. But it does not help that he is not a mathematician. Viewers of the BBC news report rudely insulted him as a "charlatan," "crank," "crackpot," and "idiot." It also does not help that elsewhere Anderson also claims to have solved the famous "mind body problem" by designing a "physical thing that is both a mind and a body."[31]

Undeterred, Anderson and his collaborators expect that transreal numbers might be used to establish a new system of computer arithmetic. But the transreals first need to be converted into a system of binary floating-point numbers, in order to be clearly compared to the IEEE 754 standard. Still, Anderson fairly points out that his system has elegant features: that all arithmetic operations on all numbers have definite numerical results and that the system is simpler compared to the standard IEEE 754 because it has a single new number nullity, rather than multiple distinct NaNs. As table 9 shows, indeed, there is simplicity in Anderson's scheme, because it does not have impossibilities, special cases, or ambiguities. Still, mathematicians have complained that the system should be useful, and thus Anderson, his collaborators, or someone else would need to make it useful in a way that is compelling to mathematicians, physicists, or others. On paper, one might choose to write $0^0 = \Phi$, but in practice a result of 1 is clearly useful in various contexts instead. And arithmetically, once Φ has emerged in a series of operations it continues to spread and absorb everything, like a sink or a black hole, because any arithmetical operation on Φ yields only Φ.

Summing up, is it true that there is no answer for division by zero? Or is it rather that we live in a time when most mathematicians still have not devised or embraced a solution that seems acceptable to them? Meanwhile, various operations with zero appear increasingly in the sciences. For one, Eli Maor turns to physics to show that the idea of division by zero has some currency. A body's density is defined as the ratio of mass to volume, $D = m/V$, and physicists believe that an immense star can collapse into a black hole such that at its center its volume becomes zero, making a division by zero, and hence that these singularities have infinite density. Maor comments: "With black holes and singularities becoming part of the physicist's daily jargon, this taboo, too, has now been broken. So perhaps the day will come when division by zero, that most sacred of mathematical taboos, will become permissible after all—subject, of course, to new rules of operation." Likewise, in statistical mechanics, physicists say that theoretically some systems can have a Kelvin temperature of infinity or more, according to this sequence of degrees from cold to hotter:

$$+0 \text{ K}, \ldots, +300 \text{ K}, \ldots +\infty \text{ K}, -\infty \text{ K}, \ldots, -300 \text{ K}, \ldots, -0 \text{ K}$$

In this context a temperature of −0 is the hottest conceivable temperature.[32] Such uses of zero and infinity are counterintuitive, but likewise, other strange and seemingly impossible quantities, such as imaginary numbers, have found abundant uses in physics. In his bestselling book *Zero: The Biography of a Dangerous Idea*, Charles Seife comments: "Zero is behind all of the big puzzles in physics. The infinite density of the black hole is a division by zero. The big bang creation from the void is a division by zero. The infinite energy of the vacuum is a division by zero. Yet dividing by zero destroys the fabric of mathematics and the framework of logic—and threatens to undermine the very basis of science." Then immediately, Seife proceeds to add: "In Pythagoras's day, before the age of zero, pure logic reigned supreme."[33] But really, pure logic did not then reign supreme, and the story of zero casts doubts on whether it does now either.

We should not assume that we're lucky enough to live in an age when all the basic operations of mathematics have been settled, when the result of division by zero in particular cannot change again. Instead, when we look at pages from old math books such as Euler's *Algebra*, we should remember that some parts of mathematics include operations and concepts involving ambiguities that admit reasonable disagreements. These are not merely aimless mistakes, but instead plausible alternative directions that mathematics has previously taken and still might take. After all, other operations that seemed impossible for centuries, such as subtracting a greater number from a lesser or taking roots of negative numbers, are nowadays perfectly fine.

7

EULER'S IMAGINARY MISTAKES

L ike zero, negative numbers have sometimes been used to derive apparent contradictions. For example, consider the following:

$$\frac{-1}{1} = \frac{1}{-1}$$

$$\sqrt{\frac{-1}{1}} = \sqrt{\frac{1}{-1}}$$

$$\frac{\sqrt{-1}}{\sqrt{1}} = \frac{\sqrt{1}}{\sqrt{-1}}$$

$$\sqrt{-1}\,\sqrt{-1} = \sqrt{1}\,\sqrt{1}$$

$$-1 = 1$$

These steps seem to prove the impossible equation, that 1 is equal to its opposite. We expect that something in the sequence of operations must be a mistake. What is it? I will give an original solution to this apparent paradox, and to do so, I'll first explain the forgotten arguments of a famous mathematician, Leonhard Euler.

The paradox above involves operations with square roots of negative numbers, the so-called imaginary numbers. While nowadays mathematicians value these numbers as being as legitimate as any others, for centuries they argued about them. These numbers seem to represent impossible operations that often led to paradoxes or contradictions, to perplexing effects, like magic. Hence, many mathematicians refused to use these numbers, and others ridiculed them with words such as "imaginary," "false," "unreal," "absurd," "nonexistent," "sophistic," "unintelligible," "merely auxiliary quantities," "impossible numbers," "quantities that exist merely in the imagination," "figments," "beings of reason," "unexecutable operations," "nonsense," "jargon," "incorrect forms," "mistaken forms," "mere algebraic forms," "expressions not susceptible of any immediate application," "hieroglyphs," "monstrous," "chimeras," "fictitious beings that cannot exist nor be understood," and so forth.[1] They even used expressions referring to "evil," "witches," and "tortures." It was not normal for mathematicians to use such nasty expressions, but one reason they did so was that they were so very sure that the notions they were criticizing were so very wrong.

Still, since it was possible to extract roots of positive numbers, equations often arose having instead negative numbers in radicals. So mathematicians struggled to make sense of these expressions. Sometimes they disagreed about whether a particular operation was possible; other times they disagreed about the results of some operations.

There is something pleasant in anecdotes about great mathematicians who made silly mistakes. Case in point: historians and mathematicians alike sometimes claim that Leonhard Euler, of all people, was confused about how to multiply imaginary numbers. His unlikely slips were published in his famous *Complete Introduction to Algebra* of 1770. As the story goes, Euler thought that the product rule

$$\sqrt{a} \times \sqrt{b} = \sqrt{(ab)} \tag{1}$$

is valid regardless of whether a and b are positive or negative. Mathematicians say that if the radical signs mean the "principal square root operation" (so that $\sqrt{4} = 2$), then Euler was wrong because his rule seems to say

$$\sqrt{-4}\,\sqrt{-9} = \sqrt{-4 \times -9} = \sqrt{36} = 6$$

whereas mathematicians now say

$$\sqrt{-4}\,\sqrt{-9} = \sqrt{-1}\,\sqrt{-1}\,\sqrt{4}\,\sqrt{9} = (-1)(2)(3) = -6$$

Again, that's if we use "principal square roots." If instead we interpret the signs to mean the "unrestricted root operation" (so that $\sqrt{4} = \pm 2$), then mathematicians say that Euler was still wrong, because

$$\sqrt{-4}\,\sqrt{-9} = \sqrt{4}\sqrt{9}(i^2) = \pm 6(-1) = \mp 6$$

which is not equal to Euler's $\sqrt{(-4 \times -9)} = \sqrt{(36)} = \pm 6$. Hence, for over two hundred years, writers have said that for negative numbers a and b the correct rule is

$$\sqrt{a} \times \sqrt{b} = -\sqrt{(ab)}. \tag{2}$$

One way or another, mathematicians say that Euler was wrong, that he was just confused or mistaken.[2] But actually, it was the mathematicians who were confused by Euler's words.

When Euler composed his *Algebra,* controversies still abounded regarding the rules of how to operate with negative and imaginary numbers. Such numbers were still often demeaned as "impossible."[3] In 1758, Francis Maseres had published his *Dissertation on the Use of the Negative Sign in Algebra,* part of his bid for the Lucasian Chair of Mathematics at Trinity College, the job that Newton once held. Maseres rejected the use of isolated negative numbers and also of imaginaries. In 1765, François Daviet de Foncenex denounced as useless the representation of imaginary numbers as constituting a line perpendicular to a line of negatives and positives.[4] And Euler himself was at the center of a dispute on the question of the logarithms of negative numbers, in opposition to Jean d'Alembert, Johann Bernoulli, and others.[5]

By 1770, the symbol i was not yet widely used to stand for $\sqrt{-1}$, though Euler had used it occasionally. Writers and typesetters used the signs $\sqrt{}$ and $\sqrt{}$ as equivalent, often meaning the *unrestricted* root operation. Nowadays, both radical signs are commonly used to indicate that only the principal (or nonnegative) root should be extracted. In what follows, the meaning of each radical will be clear from context.

Euler defined mathematics as the science of quantity, where "quantity" means whatever can increase or decrease. Hence, imaginary numbers, being neither greater nor less than zero, were generally not considered quantities.[6] The question of how to multiply square roots of negative numbers was thus one muddle among many.

Any minor defects notwithstanding, Euler's *Algebra* was hailed as being, "next to Euclid's *Elements,* the most perfect model of elementary writing, of which the scientific world is in possession."[7] Indeed, Euler's *Algebra* became the most widely read mathematics book in history, second only to the *Elements.* The *Algebra* was first published in Russian translation (two volumes, published in 1768 and 1769) before the standard German version appeared in 1770. In 1767 Euler was sixty years old and losing his eyesight. He dictated the book to a servant, a tailor's apprentice, so one might imagine that under such circumstances his account of the rules for the multiplication of roots contained simple oversights in what was otherwise a masterpiece. However, neither old age nor blindness slowed Euler's productivity or dulled his sharpness of mind, as is well known. Besides, Euler was increasingly acknowledged as the person who strikingly solved the vexing puzzle of taking logarithms of negative and complex numbers, as the latter eventually became known. So it seems stunning that he would have been confused about multiplication. A passage in his *Algebra* even seems to say that $\sqrt{-1} \times \sqrt{-4} = 2$, ridiculous though it seems. Historian Florian Cajori speculated that maybe such errors stemmed merely from typographical miscues for which Euler cannot be held accountable.[8] Others say that they involved a systematic confusion.

But surprisingly, Euler made no such mistakes. The solution to this puzzle lay buried under layers of ambiguous expressions, notations, and changing definitions. Stranger still, the history reveals defects in the rules that became incorporated into elementary algebra as we know it.

So what did Euler say about multiplying radicals? The answer is not straightforward, because *nowhere* in his *Algebra* did he even write the equation $\sqrt{a} \times \sqrt{b} = \sqrt{(ab)}$ or the equation $\sqrt{-a} \times \sqrt{-a} = \sqrt{(a^2)}$. He discussed the topic first without using the "=" sign. Rather than formulating such rules as equations, he expressed them in words and examples. Euler's *Algebra* is not a rigorous deductive treatise in which each proposition appears in its most general and exacting form. Instead, it is a textbook for students in which he introduced propositions gradually. Accordingly, we must treat his expressions with care.[9]

Instead of writing equations such as $\sqrt{-a} \times \sqrt{-b} = \sqrt{(ab)}$, Euler systematically wrote that this multiplication "gives" or "produces" $\sqrt{(ab)}$. For example: "the multiplication of $\sqrt{2}$ by $\sqrt{2}$ necessarily produces 2 . . . and in general $\sqrt{a}$ multiplied by $\sqrt{a}$ gives a."[10] For negatives too he wrote that "$\sqrt{-a}$ multiplied by $\sqrt{-a}$ gives $-a$," *instead* of writing equations.[11] This was because Euler did not define the sign = to mean "gives" or "produces." Instead, he explicitly defined it to mean "*is as much as*" and "*is equal to*."[12] Therefore, anyone who reads Euler's "gives" or "produces" as meaning equality misinterprets his meaning.

Euler specified that "from every square are given two square roots, of which one is positive, the other negative."[13] And regarding "impossible numbers," he wrote:

> The product that results when $\sqrt{-3}$ is multiplied by $\sqrt{-3}$ gives -3, so also $\sqrt{-1}$ multiplied by $\sqrt{-1}$ is -1. And in general that when we multiply $\sqrt{-a}$ by $\sqrt{-a}$, or take the square of $\sqrt{-a}$, gives $-a$. . . . Moreover, as

√a multiplied by √b gives √ab, so too will √–2 multiplied by √–3 give √6. Likewise, √–1 multiplied by √–4 gives √4, that is, 2. Thus we see that two impossible numbers, multiplied together, yield a possible or real one.[14]

Thus Euler used equation (1) to multiply imaginary numbers.

And apparently he made mistakes. Later editors of Euler's *Algebra* complained: "We should set √–2 · √–3 = √2 · √3 · (√–1)² = –√6 "[15] Likewise, mathematician Tristan Needham commented: "In 1770 the situation was still sufficiently confused that it was possible for so great a mathematician as Euler to mistakenly argue that √–2 √–3 = √6."[16] The prominent historian Ivor Grattan-Guinness also said: "Euler gave a reliable presentation; but he gaffed in his algebraic handling of complex numbers, by misapplying the product rule for square roots, √(ab) = √a√b, to write √–2√–3 = √6 instead of –√6."[17]

Moreover, Euler seemed to say that √–1 × √–4 = √4 = 2. Here too, recent famous writers on mathematics, such as Morris Kline and Paul Nahin, have said that he was mistaken, that the correct result is –2.[18] But wait, did Euler regard +2 as the *only* solution?

Euler emphasized that every square has two square roots, one positive and the other negative. He said: "This holds also for the impossible numbers, and the square root of –a is +√–a as well as –√–a."[19] He stated that this rule is *always* valid. He said: "The square root of a given number always has a double value." Likewise, in his "Researches on the Imaginary Roots of Equations," he emphasized that the quantity of imaginary roots is *always* even and *never* odd and that "by its nature the radical sign encompasses essentially the + sign as well as the – sign," that is, two solutions.[20]

In short, Euler argued that

$$\sqrt{a} \times \sqrt{b} = \sqrt{-a} \times \sqrt{-b} = \sqrt{(ab)}$$

and in particular that

$$\sqrt{a} \times \sqrt{a} = \sqrt{-a} \times \sqrt{-a} = \sqrt{(a^2)} = \pm a$$

Then mathematicians criticized his rules. Etienne Bézout, in many textbooks, gave reasons for rejecting Euler's approach. Bézout was an associate of the Paris Academy of Sciences, as well as a long-time teacher and examiner of would-be naval officers and other military personnel. In the 1781 edition of his *Course on Mathematics*, Bézout discussed the multiplication of radicals as follows. He stated rule (1), but contrary to Euler, he then asserted rule (2):

$$\sqrt{-a} \times \sqrt{-b} = \sqrt{(-a \times -b)} = -\sqrt{(ab)}$$

He explained that although every radical is susceptible to two signs, ±, a special exception happens when we multiply imaginary numbers. Bézout argued that

$$\sqrt{-a} \times \sqrt{-b} = \sqrt{a}\sqrt{-1}\sqrt{b}\sqrt{-1} = \sqrt{ab}\sqrt{(-1)^2}$$

and he explained that the root of $\sqrt{(-1)^2}$ "is not indifferently ±1," because we know that this expression comes from −1, so that the radical should undo the exponent.[21] Bézout thus rejected the general validity of (1) on the grounds that $\sqrt{-1} \times \sqrt{-1} = -1$. He argued that $(\sqrt[n]{a})^n = a$, "which is evident in general, if one realizes that the object is thus to return the quantity to its first state."[22]

Other mathematicians agreed that Euler was wrong. In Paris, Sylvestre François Lacroix agreed with Bézout. Lacroix was a renowned professor of mathematics at the École Centrale, as well as the successor to Joseph-Louis Lagrange's chair at the École Polytechnique. In his influential *Elements of Algebra,* Lacroix wrote: $\sqrt[x]{a} \times \sqrt[x]{b} = \sqrt[x]{ab}$. Yet he remarked on "certain singular cases" of that rule that could "lead to error in regard to imaginary quantities, if one does not accompany them with remarks that pertain to the properties of two terms."[23] He too said that when we don't know how the square a^2 originated, then we may give it two roots: ±a. But he praised Bézout for explaining that when we do know the origin of a^2 "it is then no longer allowed, as one returns on one's steps," to give the other root. Lacroix referred to this mistake as an "embarrassment." The textbooks of Bézout and Lacroix were widely published, revised, and reprinted for decades, including translations into German, English, Spanish, Italian, and Russian.

Their arguments convinced most mathematicians, despite some hesitations. For example, Jeremiah Day, president of Yale College, noted in his popular algebra textbook that "I have been unwilling to admit into the text rules of calculation which are commonly applied to imaginary quantities; as mathematicians have not yet settled the logic of the principles upon which these rules must be founded."[24] In particular, he said that Euler and others had asserted $\sqrt{-a} \times \sqrt{-a} = \pm a$, whereas Day argued that the result should be not +a or −a, but exclusively −a, like Bézout and Lacroix. By 1845 at least ten French editions of Bézout's *Course* had been published, and more followed. By 1868 there were twenty-two French editions of Lacroix's textbook. Other books echoed their arguments.

But in the long run, some mathematicians rejected Bézout's approach, though they agreed with its conclusion. The problem is that Bézout's approach relies on rule (1) in order to reject rule (1), which is circular. Specifically, Bézout argued that

$$\sqrt{-1}\,\sqrt{-1} = \sqrt{-1 \times -1} = \sqrt{(-1)^2}$$

but here the middle step uses the very rule he criticized, so Bézout's procedure is defective. A way to circumvent this problem was proposed soon enough, by a mathematician who edited works by both Euler and Bézout.

In an 1807 French edition of Euler's *Algebra,* Jean Garnier included a commentary criticizing Euler's rule. Garnier too was a professor of the École Polytechnique. He explained: "To multiply $\sqrt{-1}$ by $\sqrt{-1}$ is to take the square of $\sqrt{-1}$; it is therefore to return to the quantity that is under the radical. Therefore, one has $\sqrt{-1} \times \sqrt{-1} = -1$."[25] Then Garnier asserted:

$$\sqrt{a} \times \sqrt{a} = (\sqrt{a})^2 = a. \tag{3}$$

Unlike in Bézout's approach, here the exponent is *outside* the radical. Therefore, Garnier rejected rule (1) without circularity by using a distinct rule, equation (3). Still, his conclusion is the same as in Bézout and Lacroix, that the product of a radical by itself gives only one value. The conclusion spread: Euler had made embarrassing mistakes in his *Algebra.*

But did he really? Consider again the example:

$$\sqrt{-4} \times \sqrt{-9} = \sqrt{(-4 \times -9)}$$

following Euler's claims. Contrary to the arguments of Bézout and Lacroix, we may disregard any idea that this equation's right side "comes from" the left side, because we can just as well say the opposite. We can think about the equation without any temporal sequence, of one side being prior. Simply, two expressions given simultaneously are separated by the equality sign, and we want to know whether they are really equivalent. To determine this, we simplify each expression directly. Take the right side, $\sqrt{(-4 \times -9)}$: by multiplying first, we get $\sqrt{36}$, after which the unrestricted radical yields ±6. Now take the left side of the equation, $\sqrt{-4} \times \sqrt{-9}$. We extract the square roots, $\pm2i \times \pm3i$, where the term $\pm2i$ designates two imaginary roots for $\sqrt{-4}$ and $\pm3i$ designates two roots for $\sqrt{-9}$. A pair of double signs was used systematically by Euler to represent four values.[26] So we have

$$(+2i) \times (+3i) = -6, \quad (+2i) \times (-3i) = +6, \quad (-2i) \times (+3i) = +6, \quad (-2i) \times (-3i) = -6$$

These results are summarized by writing $(\pm2i) \times (\pm3i) = \pm6$, or by another way of expressing multiplication of four roots.[27] Therefore, we finally have

$$\sqrt{-4} \times \sqrt{-9} = \sqrt{(-4 \times -9)}$$
$$\pm 2i \times \pm 3i = \sqrt{36}$$
$$\pm 6 = \pm 6$$

This same procedure yields the same result if the numbers involved are positive. In this way, surprisingly, the equation $\sqrt{a} \times \sqrt{b} = \sqrt{(ab)}$ is valid for both positive and negative numbers. In short, Euler's approach works!

His approach is coherent because by admitting both values of each radical and by using the fourfold multiplication of signs, equation (1) gives the same results when applied to any pair of negative numbers as it does for the corresponding pair of positives. It also works for any combination of negatives and positives.

Moreover, the same procedure explains Euler's statements on *dividing* imaginaries, which *also* have been criticized as erroneous by historians such as Cajori and Grattan-Guinness.[28] To show this, let's return to the paradox at the start of this chapter. By first assuming that $-1 = -1$, we seem to derive the impossibility that $-1 = 1$. Mathematicians and teachers explain that the problem, the fallacy, is that the argument involves an illegitimate operation, that root extraction is distributed over division:

$$\sqrt{\frac{-1}{1}} = \sqrt{\frac{1}{-1}}$$
$$\frac{\sqrt{-1}}{\sqrt{1}} = \frac{\sqrt{1}}{\sqrt{-1}}$$

The problem, teachers say, is that we cannot assume that the rules of real numbers apply to imaginaries and therefore that we cannot distribute radicals over division of negative numbers. Division is often defined in terms of multiplication; therefore, by requiring that rule (1) is not valid for negative numbers, the step taken in the equations above is forbidden.

By contrast, here's the new solution I promised, and here's how Euler's approach solves the apparent paradox. We now have

$$-1 = -1$$
$$\frac{-1}{1} = \frac{1}{-1}$$
$$\sqrt{\frac{-1}{1}} = \sqrt{\frac{1}{-1}}$$
$$\frac{\sqrt{-1}}{\sqrt{1}} = \frac{\sqrt{1}}{\sqrt{-1}}$$
$$\sqrt{-1}\sqrt{1}\,\frac{\sqrt{-1}}{\sqrt{1}} = \frac{\sqrt{1}}{\sqrt{-1}}\,\sqrt{-1}\sqrt{1}$$

$$\sqrt{-1}\sqrt{-1} = \sqrt{1}\sqrt{1}$$
$$\sqrt{-1\times-1} = \sqrt{1\times1}$$
$$\sqrt{1} = \sqrt{1}$$
$$\pm1 = \pm1$$

There is no contradiction! Here, the operations that apply to real numbers apply likewise to imaginary numbers, and exceptions and restrictions are not needed, in contrast to the standard approach.

The main difference between the two approaches is the question of how we multiply radicals. By first solving each radical, we can state an equivalent issue: how do we multiply plus or minus signs? One option is to write

$$\sqrt{1}\times\sqrt{1} = \pm1 \times \pm1 = \begin{bmatrix} +1\times+1 = +1 \\ +1\times-1 = -1 \\ -1\times+1 = -1 \\ -1\times-1 = +1 \end{bmatrix} = \pm1$$

This is just another way to explain Euler's approach. This same procedure can be written algebraically:

$$\left(\sqrt{a}\right)^2 = \left(\pm r\right)^2 = \left(\pm r\right)\times\left(\pm r\right) = \begin{bmatrix} (+r)\times(+r) = +a \\ (+r)\times(-r) = -a \\ (-r)\times(+r) = -a \\ (-r)\times(-r) = +a \end{bmatrix} = \pm a$$

Or instead, we can consider a different rule for multiplying double signs:

$$\sqrt{a}\times\sqrt{a} = \left(\sqrt{a}\right)^2 = \left(\pm r\right)^2 = \begin{bmatrix} (+r)^2 = (+r)\times(+r) = +a \\ or \\ (-r)^2 = (-r)\times(-r) = +a \end{bmatrix} = +a$$

Here we multiply each square root only by itself. This is what Garnier explicitly argued in his note to Euler's text:

$$(\pm r)^{2m} = \left((\pm r)^2\right)^m = (+r^2)^m = +\,r^{2m}$$

where the result is only the positive solution.[29] So, depending on how we define the operation of squaring the ± sign, we get different results. Some people might view one rule as "more natural" than the other. Thus, although many people viewed $\sqrt{-1}\times\sqrt{-1} = -1$ as necessarily true, some mathematicians later regarded this as just a useful "convention" or "supposition." For example, this is how professors Isaac Todhunter and Charles Smith, at Cambridge, described this equation.[30] They had studied the role of conventions in the foundations of algebra, thanks to the works of other British mathematicians. In mathematics, certain rules are not discovered; they are invented, like rules in a game.

Mathematicians chose to follow Garnier, partly because they did not understand Euler. Whichever rules are chosen, however, have significant consequences. For example, following the rules of Bézout and Garnier, we have

$$\sqrt{(4)^2} = \sqrt{16} = 4, \text{ and } \left(\sqrt{4}\right)^2 = (\pm 2)^2 = 4$$

By contrast, following Euler's rules, we would have

$$\sqrt{(4)^2} = \sqrt{16} = \pm 4, \text{ and } \left(\sqrt{4}\right)^2 = (\pm 2)^2 = \pm 4$$

Interestingly, both of these different approaches, with different results, can be summarized by the same algebraic equation:

$$\sqrt[n]{(a)^m} = \left(\sqrt[n]{a}\right)^m$$

But just as there were disagreements in numerical results, some mathematicians also disagreed about algebraic equations. For example, Charles Smith, master of Sidney Sussex College of Cambridge University, expected that instead of the algebraic rule above, we can sometimes have

$$\sqrt[n]{(a)^m} \neq \left(\sqrt[n]{a}\right)^m$$

Smith argued: "It should be remarked that it is not strictly true that $\sqrt[n]{(a^m)} = \left(\sqrt[n]{a}\right)^m$ . . . unless by the nth root of a quantity is meant only the *arithmetical* root. For example, $\sqrt[2]{(a^4)}$ has *two* values, namely $\pm a^2$, whereas $\left(\sqrt[2]{a}\right)^4$ only the value $+a^2$."[31] Thus, Smith required that

$$\sqrt{(4)^2} = \sqrt{16} = \pm 4, \text{ but } \left(\sqrt{4}\right)^2 = (\pm 2)^2 = 4$$

The point is that mathematicians were tailoring and choosing whichever rules seemed more reasonable to them, and each choice came at a price. In some approaches, radicals and powers became inverse operations, while in other approaches they were not exactly inverse. In some approaches, a rule for positive numbers became valid for negative numbers too, while in other approaches they did not obey the same rules. In some approaches, some algebraic relations were valid for all cases, all numbers, while in others there were special cases in which such rules were not valid.

Return now to the main reason why mathematicians thought that Euler was wrong. They rejected the general validity of his product rule by requiring that $\sqrt{a} \times \sqrt{a} = a$, which is a convenient and elegant rule. Also, by rejecting Euler's rules, we can conveniently write: $a^{\frac{1}{2}} \times a^{\frac{1}{2}} = a^{\frac{1}{2}+\frac{1}{2}}$ Thus, the principle of exponents, that $a^b + a^c = a^{b+c}$, is not always true in Euler's algebra.

However, Euler's approach has a beautiful advantage over those of Bézout, Lacroix, and others. Only in Euler's approach does the following principle apply universally: equal operations performed on both sides of an equation always preserve the equality. For example, consider the identity

$$(+a)^2 = (-a)^2$$

Now extract square roots on both sides. If we respect Euler's rule that every square root radical has two values, then

$$\sqrt{(+a)^2} = \sqrt{(-a)^2}$$

$$\pm a = \pm a$$

Otherwise, according to the rule of Bézout and Lacroix, we get

$$\sqrt{(+a)^2} = \sqrt{(-a)^2}$$

$$+a = -a$$

which is a clear contradiction, unless $a = 0$. Thus, Euler's approach was rejected in favor of an approach that violated one of the most fundamental rules of arithmetic: *that equal operations on both sides of an equality preserve the equality.* In this light, it was not Euler who was just wrong.

Should all the rules of arithmetic hold in algebra? Should all numbers follow the same rules? Rather than using a single general rule, mathematicians used various rules to multiply radicals. Some, such as Lacroix, asserted Bézout's rule. Later many others, such as George Peacock, adopted Garnier's rule instead.[32] Euler's approach was a third alternative, the only one that respects the rule that *every* nonzero square root has two values.

The reactions against Euler's approach stemmed partly from notions that today are considered to be nonmathematical. In particular, algebra was often viewed as involving relations in time. Some philosophers, such as Kant, intimately associated notions of numerical order with temporal order. Likewise, mathematician William Rowan Hamilton construed algebra as the "Science of Pure Time," just as geometry was viewed as the science of space.[33] Alongside such overt claims, some mathematical words involved temporal notions. For example, terms such as *root* and *product* seemed to presuppose temporal ordering. One side of an equation was sometimes said to precede the other. Knowledge of what came from where served to decide the acceptability of some solutions.

Mathematicians gradually abandoned those perspectives. Yet there remained rules that had been introduced partly on the basis of such notions, such as Bézout's rule. If only Euler's product rule had been properly understood, it could have been appreciated, because it preserves arithmetical relations. But that's hypothetical; the fact is that mathematicians increasingly abandoned the idea that algebra must conform to arithmetic, just as they had earlier rejected physical analogies as justifications for algebraic rules.

Most algebraists did not grasp Euler's approach because it clashed with other rules that they posited. In the end, it comes down to choices of axioms. If we assume that all square roots have two values *and* require the fourfold multiplication of double signs, then Euler's results are justifiable. Otherwise, the product rule can be restricted by positing independent rules. This restriction trades economy and generality of axioms for the convenience of simpler results. It has the advantage of reducing the proliferation of the ambiguous ± sign. However, Euler's approach has advantages. It ensures the commutativity of the unrestricted radical and squaring operations. It admits into algebra certain general properties, such as $\sqrt{a}\sqrt{b} = \sqrt{(ab)}$, and $(\sqrt{a})/(\sqrt{b}) = \sqrt{(a/b)}$, that are otherwise restricted.

Before a standard system of laws was adopted, mathematicians had some freedom to choose whatever rules they saw fit, and thus they developed algebra in different directions. Euler's account of multiplication was rejected because it was not clearly understood and also because it clashed with properties that mathematicians preferred. Over time, the expression $\sqrt{-1}$ served increasingly to signify the single numerical value i, rather than $\pm i$, while the $\sqrt{}$ sign became used more and more to designate only nonnegative roots. Such conventions simplified algebra, eliminating multiple solutions that complicate some calculations.[34] But still, Euler's alternative approach shows that even the elementary rules of algebra admit variations that lead to symmetric and elegant results.

8

THE FOUR OF PYTHAGORAS

It is well known that any complex number, such as 4 + 2*i*, can be represented by a point or a line in a plane. The real and imaginary parts of this number correspond to *x* and *y* coordinates on the so-called complex plane, as illustrated in the figure.

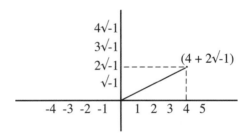

Some teachers love this: it seems to clearly give meaning to complex numbers by connecting numbers and geometry: every single number, real or complex, corresponds uniquely to a single point in a plane. If we take this sheet of paper, this page, as representing the complex plane, then the period at the end of this sentence corresponds to a single complex number. Neat. But there is a huge problem: space does not really correspond in a one-to-one way with complex numbers. If the present page represents the complex plane, and every position on the page corresponds to a single number, then what is the number that corresponds to the very tip of your nose? What numbers correspond to any of the points *outside* the plane?

The apparent direct match between numbers and geometry is an illusion, a simplification produced by disregarding most of geometry by restricting attention to only one plane, an infinitesimally thin slice of space. This immense mismatch between geometry and numbers led a few mathematicians to wonder: if complex numbers correspond to positions in two-dimensional space, might there exist unknown numbers that correspond to positions in three-dimensional space? Since mathematicians were unaware of imaginary numbers for thousands of years, might there not exist some stranger numbers that had waited even longer to be discovered?

William Rowan Hamilton was a mathematician who eventually wrestled with these questions. Hamilton grew up in Ireland, with an uncle who helped him learn many languages: Latin, Greek, Hebrew, Italian, French, Persian, Arabic, Hindustani, Sanskrit, Syriac, Marathi, and Malay. As a young man, he studied classics and science at Trinity College, Dublin. He was especially good at mathematics. When Hamilton was barely twenty-two years old, in 1827, shortly before he graduated, several professors were so impressed by him that they appointed him to a chair in astronomy. He did not want to become an astronomer, but he accepted the honor because the professorship would allow him to pursue other interests freely.

Also in 1827, Hamilton met a famous English poet, William Wordsworth. Hamilton was delighted because he hoped to pursue poetry, alongside science and mathematics. As a young man, he had experienced the "charm severe" of mathematics that Wordsworth described in his poem "The Excursion":

> His Step-father supplied; books that explain
> The purer elements of truth involved
> In lines and numbers, and, by charm severe,
> (Especially perceived where nature droops
> And feeling is suppressed), preserve the mind
> Busy in solitude and poverty.
> These occupations oftentimes deceived
> The listless hours, while in the hollow vale,
> Hollow and green, he lay on the green turf
> in pensive idleness.[1]

Caught between mathematics, science, and poetry, Hamilton began to struggle with the problem of space. If the complex number

$$x + iy$$

represents a point, it can also be used to mean a line on the plane, by making the line's other endpoint be point 0. But if so, what kind of number would describe a line in real space, say, in three dimensions? Might there exist some sort of a three-part number, a "triplet"? If perhaps we write

$x + iy + jz$

could this expression represent a line in three-dimensional space? But if so, what is j? Is it imaginary? What happens when we carry out arithmetical operations on this number? Hamilton struggled to imagine rules for such unknown triple numbers.

In 1829, the poet Wordsworth visited Hamilton at the Dunsink observatory. In friendly conversations, Hamilton mentioned that he disliked a disparaging passage that Wordsworth had written about science in "The Excursion":

> Science then
> Shall be a precious Visitant; and then,
> And only then, be worthy of her name.
> For then her Heart shall kindle; her dull Eye,
> Dull and inanimate, no more shall hang
> Chained to its object by brute slavery;
> But taught with patient interest to watch
> The processes of things, and serve the cause
> Of order and distinctness, not for this
> Shall it forget that its most noble use,
> Its most illustrious province, must be found
> In furnishing clear guidance, a support
> Not treacherous, to the Mind's *excursive* Power.[2]

Hamilton complained that Wordsworth did not convey enough reverence for science. The veteran poet replied that he revered science inasmuch as it aimed to elevate the mind to contemplate God's works—but that some scientists lacked this aim, put it out of view, and hence their science actually *degraded* humans as it sought merely to collect facts for their own sake or to apply them to material uses, excluding the imagination. Wordsworth said that he criticized not the heroes of science, such as Newton, but only its army of soldiers who lack feeling, enthusiasm, or hope.

Hamilton replied that "*Intellectual*" faculties share at least equal rank with "*Imaginative*" faculties. Yet he believed that the objects of mathematics are not imaginary. Hamilton said that mathematics links men to beings of a higher nature, that circles and triangles really exist in minds and in the nature of things, that they are not arbitrary symbols created by human invention. Wordsworth smiled but replied that this reminded him of the Platonic doctrine that beautiful forms exist, for example, inside marble, *before* a sculptor uncovers them.[3]

In further exchanges, Hamilton shared some of his amateur poems with Wordsworth. The veteran poet was not impressed. With sympathy, he convinced Hamilton that he lacked the dedication to make great poetry. He explained that poetry is a demanding art that requires not just inspiration but continuous and painstaking efforts, "that Poetry alike and Science are Muses that refuse to be successfully wooed by the same suitor."[4] Hamilton concluded that he should painfully say farewell to poetry, then focus on science and mathematics.

Hamilton continued to struggle with the problem of triplets in space, and he continued his teaching duties. In introductory lectures on astronomy, he told his students about the hopeful inner life of a mathematician:

> These purely mathematical sciences of algebra and geometry are sciences of the pure reason, deriving no weight and no assistance from experiment, and isolated, or at least isolable from all outward phenomena. The idea of order, with its subordinate ideas of number and of figure, we must not indeed call innate ideas, if that phrase be defined to imply that all men must possess them with equal clearness and fullness; they are, however, ideas which seem to be so far born with us, that the possession of them in any conceivable degree, appears to be only the development of our original powers, the unfolding of our proper humanity. Foreign, in so far as they touch not the will, nor otherwise than indirectly influence our moral being, they yet compose the scenery of an inner world, which depends not for its existence on the fleeting things of sense, and in which the reason, and even the affections, may at times find a home and a refuge. The mathematician, dwelling in that inner world, has hopes, and fears, and vicissitudes of feeling of his own; and even if he be not disturbed by anxious yearnings for an immortality of fame, yet has he often joy, and pain, and ardour: the ardour of successful research, the pain of disappointed conjecture, and the joy that is felt in the dawning of a new idea. And when, as on this earth of ours must sometimes happen, he has sent forth his wishes and hopes from that lonely ark, and they return to him, having found no resting place:

while he drifts along the turbulent current of passion, and is tossed by the storm and glory of grief, some sunny bursts may visit him, some moments of delightful calm may be his, when his old habits of thought recur, and the "charm severe" of lines and numbers is felt at intervals again.[5]

Hamilton's passion for mathematics was accompanied by annoyance at its foundations. In 1835, he published a booklet in which he complained about the seemingly senseless rules of negative and imaginary numbers:

> It requires no peculiar scepticism to doubt, or even to disbelieve, the doctrine of Negatives and Imaginaries, when set forth (as it has commonly been) with principles like these: that a *greater magnitude may be subtracted from a less,* and that the remainder is *less than nothing;* that *two negative numbers,* or numbers denoting magnitudes each less than nothing, may be *multiplied* the one by the other, and that the product will be a *positive* number, or a number denoting a magnitude greater than nothing; and that although the *square* of a number, or the product obtained by multiplying that number by itself, is therefore *always positive,* whether the number be positive or negative, yet that numbers, called *imaginary,* can be found or conceived or determined, and operated on by all the rules of positive and negative numbers, as if they were subject to those rules, *although they have negative squares,* and must therefore be supposed to be themselves neither positive nor negative, nor yet null numbers, so that the magnitudes which they are supposed to denote can neither be greater than nothing, nor less than nothing, nor even equal to nothing. It must be hard to found a SCIENCE on such grounds as these.[6]

Hamilton then demonstrated that instead of accepting the usual theory of imaginary numbers, he could reproduce their operations without using imaginary numbers at all. Instead, he showed that complex numbers such as $x + iy$ could be replaced with ordered pairs of real numbers: algebraic couples (a, b). He showed how to rigorously replicate all the properties of complex numbers without using imaginary numbers at all.

Still, after many years of obsessive work, Hamilton had not solved the problem of space and numbers: can there exist triplets such as $x + iy + jz$? As with i, he supposed that the new imaginary term j would also have the property that $j^2 = -1$. He hoped that the triplets would serve to analyze lines in space. He expected that operations on triplets would produce other triplets (just as operations with complex numbers produce other complex numbers). Easily, adding and subtracting triplets produced other triplets. For example:

$$(3 + 2i + 5j) + (1 + 3i + 4j) = 4 + 5i + 9j$$

Likewise, here's an example of subtracting triplets:

$$(4 + 5i + 9j) - (3 + 2i + 5j) = 1 + 3i + 4j$$

But multiplication was problematic.[7] For example, if we try to multiply the triplet $x + iy + jz$ by itself, we get

$$x^2 + xiy + xjz + iyx + (-y^2) + ijyz + jzx + jiyz + (-z^2)$$

which simplifies to

$$x^2 - y^2 - z^2 + 2ixy + 2jzx + 2ijyz$$

Is this a triplet? Its first three terms are real numbers (positive or negative or zero) and thus their sum is a real number, as in the first term of a triplet. Next, $2ixy$ corresponds also to the usual kind of imaginary term i, and the $2jzx$ corresponds to Hamilton's new imaginary j. But what about the last term: $2ijyz$? What was it? If it were absent, then it would be reasonable to say that the multiplication of triplets produced another triplet. So how might it be erased? Hamilton saw that one way to erase it was to assume that

$$ij = 0$$

Or instead, he realized that another way to erase it would be to assume that

$$ij = -ji$$

In that case, in the sum of nine terms above, the terms $ijyz$ and $jiyz$ would cancel out, eliminating the ambiguous term $2ijyz$. But the latter alternative involves the bizarre implication that the commutative law of multiplication, $ab = ba$, would not be valid for $i \times j$. This commutative law *had applied to all numbers*, even i, so why would it not apply to j? Nevertheless, Hamilton pondered the two ways of defining the multiplication of a triplet by itself to produce another triplet.

So far, so good, but there were other aspects of multiplication that did not work. When we multiply complex numbers, there are some convenient properties that Hamilton hoped would also apply when multiplying triplets. To check whether triplets behave like complex numbers, we first have to explain properties of complex multiplication. If we multiply two arbitrary complex numbers, we get

$$(a + bi)(x + yi) = (ax - by) + (ay + bx)i$$

Since the right side of the equation has only one imaginary term, we can summarize the equation by writing

$$(a + bi)(x + yi) = (e + fi)$$

Now, by disregarding the imaginary terms to look at relations among real numbers, mathematicians had found that the following holds generally:

$$(a^2 + b^2)(x^2 + y^2) = (e^2 + f^2)$$

This numerical relation had become known as the "law of norms." Mathematicians had found a clear geometric significance among these quantities: If x and y are the horizontal and vertical components of a line represented by the complex number $x + yi$, then, using the so-called Pythagorean theorem, the *length* of that complex line segment is given by $\sqrt{x^2 + y^2}$. Thus, if we take the square roots of both sides of the equation above, we get

$$\sqrt{(a^2 + b^2)(x^2 + y^2)} = \sqrt{(e^2 + f^2)}$$

which, using the product rule (as discussed in the previous chapter), yields the equation that has geometric meaning:

$$\sqrt{a^2 + b^2}\sqrt{x^2 + y^2} = \sqrt{e^2 + f^2}$$

This says that, on the right side, the length $\sqrt{e^2 + f^2}$ of the product of two complex numbers ($a + bi$ multiplied by $x + yi$) is equal to, on the left side, the product of the lengths of the two line segments.

So, interested in the geometric meaning of algebraic expressions, Hamilton wanted to know whether the law of norms was valid for triplets. Multiplying a triplet by itself, again, we get

$$(x + iy + jz)(x + iy + jz) = (x^2 - y^2 - z^2) + (2ixy) + (2jxz) + (2ijyz)$$

If we set $ij = 0$, then delete the remaining imaginary terms (as we did for the norm of complex numbers), and take the square of x, y, z, we have:

$$(x^2 + y^2 + z^2)(x^2 + y^2 + z^2) = (x^2 - y^2 - z^2)^2 + (2xy)^2 + (2xz)^2$$

Hamilton found that this equation is numerically valid, creating a fair analogy to the usual law of norms.

But so far we have only considered the multiplication of a triplet by itself—would multiplication work in the same way when we multiply any two arbitrary triplets? Here Hamilton encountered problems. If we multiply, for example, $(a + ib + jc)(x + iy + jz)$, we get

$$(ax - by - cz) + i(ay + bx) + j(az + cx) + (ijbz + jicy)$$

Is this a triplet? No, because it has four terms, but if we set $ij = ji = 0$, then we have

$$(ax - by - cz) + i\,(ay + bx) + j\,(az + cx)$$

This result seems to work, but unfortunately, it leads to a problem with the law of norms. If we disregard the i and j terms, as before, and square each of the three terms, we obtain

$$(a^2 + b^2 + c^2)\,(x^2 + y^2 + z^2) = (ax - by - cz)^2 + (ay + bx)^2 + (az + cx)^2$$

But this equation does not work: it's numerically false. For example, if we set $a = 1$, $b = 2$, $c = 3$, $x = 4$, $y = 5$, $z = 6$, we get

$$(a^2 + b^2 + c^2)\,(x^2 + y^2 + z^2) = 1{,}078$$

whereas, for the other side, we get

$$(ax - by - cz)^2 + (ay + bx)^2 + (az + cx)^2 = 1{,}069$$

The law of norms fails to hold for these triplets. The two results are not equal. But the difference between them, $1078 - 1069 = 9$, is noteworthy, as we'll see.

Summing up, when Hamilton multiplied arbitrary triplets he did not get a triplet, and when he deleted the fourth term, by setting $ij = 0$, the results violated the law of norms, and the product of lengths was therefore geometrically confusing.

But Hamilton realized that there was another alternative: to suppose instead that $ij = -ji$. Consider again the product of two triplets, the four-part expression

$$(ax - by - cz) + i\,(ay + bx) + j\,(az + cx) + (ijbz + jicy)$$

By making $ij = -ji$, we now have

$$(ax - by - cz) + i\,(ay + bx) + j\,(az + cx) + ij\,(bz - cy)$$

If we convert this expression, as before, to match the law of norms, we find the following value for the term that we had eliminated previously, the troublesome fourth term, if we had kept it and squared it like the others:

$$(bz - cy)^2 = (2 \times 6 - 3 \times 5)^2 = 9$$

This is the quantity that was missing above! In this case, it is numerically true that

$$(a^2 + b^2 + c^2)\,(x^2 + y^2 + z^2) = (ax - by - cz)^2 + (ay + bx)^2 + (az + cx)^2$$
$$+ (bz - cy)^2$$

So Hamilton found that the law of norms works perfectly under the assumption that $ij = -ji$.

Problem solved? Not at all, because if he made this assumption, then the multiplication of two triplets does not make a triplet, but instead a four-part quantity, which seems geometrically meaningless: it is not a line segment in three-dimensional space.

Hamilton expected that the multiplication of two arbitrary lines would produce a line having a length that is the numerical product of the lengths of the two lines. But this did not work for triplets that were not in the same plane; instead, there appeared an ambiguous fourth term. How to eliminate it? Hamilton reconsidered his makeshift rules, his assumptions. Was there any way to modify them to erase the superfluous term?

Hamilton's daily life was complicated by his dysfunctional relationship with his wife, Helen. They had two small sons and a daughter, but Helen was often ill and spent months away from their home, also caring for her own ailing mother. Hamilton drank much wine, frequently; he became an alcoholic. And he was often depressed, yet he obsessively persisted in mathematics, working especially on the problem of triplets.

His study room was a mess. The floor and furniture were littered by disorganized stacks of paper, covered in algebraic scribbles. Unfinished dishes sat with desiccated food, under piles of paper. He compulsively wrote equations; when he did not have paper, he wrote wherever he could, sometimes on his fingernails, even on eggshells.

In the fall of 1843, he was thirty-eight years old. He struggled every day to multiply his imaginary triplets. He told his two young sons about this mathematical problem. And in the mornings, on coming down to breakfast, they repeatedly asked him: "Well, Papa, can you *multiply* triplets?" And in turn, he "was always obliged to reply, with a sad shake of the head: 'No, I can only *add* and subtract them.'"[8]

Then one fall day, early in the evening, Hamilton and his wife were walking in the northwest outskirts of Dublin, toward the city, along the narrow Royal Canal. Hamilton later explained that as they walked, although Helen "talked with me now and then, yet an *under-current* of thought was going on in my mind."[9] As they approached a small stone bridge, Broome Bridge on Broombridge road, which crosses the canal, toward the city, Hamilton thought about his seemingly impossible three-part numbers. What if instead of using only three numbers, x, i, j, he used instead *four* numbers: x, i, j, k? Hamilton later recalled:

> I then and there felt the galvanic circuit of thought *close;* and the sparks which fell from it were the *fundamental equations between i, j, k,* *exactly such* as I have used them ever since. I pulled out, on the spot,

a pocket-book, which still exists, and made an entry, on which, *at the very moment*, I felt that it might be worth my while to expend the labour of at least ten (or it might be fifteen) years to come. But then it is fair to say that this was because I felt a *problem* to have been at the moment *solved*—an intellectual want relieved—which had *haunted* me for at least *fifteen years before*.[10]

He scribbled into that pocket-book, which Helen had given him in 1840.[11]

$$i^2 = j^2 = k^2 = -1.$$
$$ij = k \quad jk = i \quad ki = j,$$
$$ji = -k \quad kj = -i \quad ik = -j.$$

It was Monday evening, 16 October 1843, and Hamilton sensed the historic significance of his breakthrough. Right then, in his excitement, Hamilton stepped toward Broome bridge ("Brougham," as he mistakenly spelled it) and took out his pocketknife:

At last a *result,* whereof it is not too much to say that I felt *at once* the importance. An *electric* circuit seemed to *close;* and a spark flashed forth, the herald (as I *foresaw, immediately*) of many long years to come of definitely directed thought and work, by *myself* if spared, and at all events on the part of *others,* if I should even be allowed to live long enough distinctly to communicate the discovery. Nor could I resist the impulse—unphilosophical as it may have been—to cut with a knife on a stone of Brougham Bridge, as we passed it, the fundamental formula with the symbols, *i, j, k;* namely,

$$i^2 = j^2 = k^2 = ijk = -1$$

which contains the Solution of the Problem.[12]

Regarding multiplication, Hamilton realized that without erasing *ij*, as he had previously done, multiplication would work. Thus he suddenly posited that the product of *i* and *j* should be a *new* kind of imaginary number: *k*. His new "numbers" did not consist of three parts but of four, so Hamilton named them "Quaternions." Here is one such number, for example:

$$3 + 2i + 16j + 9k$$

Such four-part numbers, multiplied, produce other four-part numbers; all having the form:

$$v + ix + jy + kz$$

To Hamilton, such interesting expressions seemed to refer to a fourth dimension. Moreover, these numbers involved a perplexing property: they disobeyed a traditional, fundamental property of multiplication. Mathematicians believed that the product of *any* two numbers is the same *regardless* of the order in which they are multiplied:

$$ab = ba$$

But now, Hamilton violated the commutative rule by requiring that

$$ij = -ji \neq ji$$

On that same Monday night, conscious of the historic significance of his ideas, Hamilton told some colleagues about his realization and began writing. Referring to multiplication, he noted: "If the factor lines be perpendicular to each other, the product line, being still perpendicular to both, is in length = the product of their lengths." He also speculated that quaternions might become useful for the analysis of physical and mathematical problems, involving electricity, polarities, intensities, and spherical trigonometry; for example, that "in the quaternion (v, x, y, z), xyz may determine *direction and intensity;* while v may determine the *quantity* of some agent such as electricity."[13]

The next day he wrote to his friend Robert Graves. Hamilton admitted: "The train of thought is curious, almost wild, but I believe that the mathematical chain has kept the wings of fancy from soaring altogether out of bounds—though a fourth dimension of space is doubtless something like that step."[14] Soon, Robert's brother the mathematician John Graves replied in various letters, praising Hamilton's quaternions and pondering their properties. In one letter, John Graves voiced a puzzling concern:

> There is still something in the system which gravels me. I have not yet any clear views as to the extent to which we are at liberty arbitrarily to create imaginaries, and to endow them with supernatural properties. You are certainly justified by the event. You have got an instrument that facilitates the working of trigonometrical theorems and suggests new ones, and it seems hard to ask for more; but I am glad that you have glimpses of physical analogies. But supposing that your symbols have their physical antitypes, which might have led to your quaternions, what right have you to such luck, getting at your system by such an *inventive* mode as yours? If with your Alchemy you can make three pounds of gold, why should you stop there?[15]

Indeed! If he could invent new imaginary numbers, and hence new mathematics, why not invent others?

Hamilton's definitions of i, j, k entailed that

$$\sqrt{-1} = i, \ \sqrt{-1} = j, \ \sqrt{-1} = k$$

So it would now seem that the square root of -1 is not just one thing, but three. No, remember that $-i \times -i = -1$, so we also have

$$\sqrt{-1} = -i, \ \sqrt{-1} = -j, \ \sqrt{-1} = -k$$

Are there *six* values for the square root of -1? A plausible objection might be that quaternions are self-contradictory because these equations imply

$$\sqrt{-1} = i = j = k$$

whereas Hamilton established that $i \neq j \neq k$. However, this objection does not work because of rules that mathematicians had previously accepted. They had already accepted that

$$\sqrt{-1} = i \neq -i = \sqrt{-1}$$

Mathematicians had accepted that some operations have multiple solutions. For example, consider the cube roots of 1:

$$\sqrt[3]{1} = 1, \quad \sqrt[3]{1} = \frac{-1 + \sqrt{-3}}{2}, \quad \sqrt[3]{1} = \frac{-1 - \sqrt{-3}}{2}$$

Mathematicians *avoided* contradictions with cube roots by not writing

$$1 = \frac{-1 + \sqrt{-3}}{2}$$

although both sides of the equation are equal to $\sqrt[3]{1}$. Therefore, since mathematicians allowed multiple solutions for particular equations, they could hardly complain that Hamilton's quaternions involved contradictions because of the multiple values for the square roots of -1.

So, regarding the new imaginary numbers, John Graves asked, if we can make three pounds of gold, why stop there? Why not proceed to design new numbers, new mathematics? Consequently, Graves analyzed whether perhaps there might exist imaginary numbers with more parts than quaternions. Then he promptly made a system of eight-part numbers "octaves," such as

$$v + ai + bj + ck + dl + em + fn + go$$

where l, m, n, o, are new imaginary numbers. Again, they are each different from one another, but the square of each is equal to -1. He showed that the multiplication of two sums of eight perfect squares produces another sum of eight perfect squares.[16] On 26 December 1843, John Graves sent a draft of his scheme to Hamilton, but Hamilton raised a few objections. He complained that Graves's numbers disobeyed the associative law of multiplication:

$$a \, (bc) = (ab) \, c$$

But if Hamilton had broken the commutative law, why not also break the associative law? Yet his complaints led Graves to abstain from publishing promptly, and thus he lost that priority, because the mathematician Arthur Cayley devised the same kind of system and published it first, in 1845.[17] Therefore, the octaves or "octonions" became known as "Cayley numbers." Still, John Graves published his scheme soon afterward. His brother Charles Graves also published another new mathematical system, in 1846. And Augustus De Morgan, who also received early notice of Hamilton's work, published five new numerical systems.

So, soon after Hamilton announced his breakthrough, other mathematicians devised new number systems. Were they inventing new mathematics? Or were they unveiling eternal structures from within the unseen nature of things? Did the sculptures preexist inside the marble?

Ideas of four- or eight-part imaginary numbers might seem invented. But Hamilton did not believe that his quaternions were an invention. Instead, he felt that he had discovered natural mathematical things, which could have been discovered by the ancients. He conjectured that the four terms of quaternions correspond to four dimensions: three of space and one of time. They seemed to promise a natural mathematics ideally suited for physics. In 1846, inspired by quaternions, Hamilton composed the following sonnet at the Dunsink Observatory:

THE TETRACTYS

Or high Mathésis, with its "charm severe
Of line and number," was our theme; and we
Sought to behold its unborn progeny,
And thrones reserved in Truth's celestial sphere;
While views before attained became more clear:
And how the One of Time, of Space the Three,
Might in the Chain of Symbol girdled be:
And when my eager and reverted ear
Caught some faint echoes of an ancient strain,
Some shadowy outline of old thoughts sublime,

Gently He smiled to mark revive again,
In later age, and occidental clime,
A dimly traced Pythagorean lore,
A westward floating, mystic dream of FOUR.[18]

The strange title word, *Tetractys,* is a Greek word corresponding to the Latin word *Quaternio.* The title refers to a mystical oath or number that supposedly entranced the cult of Pythagoras. Early in the second century CE, Theon of Smyrna recorded a mystical oath: "I swear by he who transmitted the quaternion into our souls, the source of eternal nature." Theon claimed that the one who had transmitted this knowledge was Pythagoras, because "that which has been said about the Tetractys in effect seems to come from his philosophy."[19]

This oath was repeated by the very late biographers of Pythagoras, namely Porphyry and Iamblichus, and it was also echoed in "The Golden Verses," traditionally attributed to Pythagoras without evidence: "I swear it by him who has transmitted into our Soul the sacred Quaternion, The Source of Nature, whose Course is Eternal." [20] Another ancient text stated: "'By him who transmitted to our soul the Tetraktys, which has the spring and root of ever-flowing nature.' And our soul, he says, is composed of the Tetrad, for it is intelligence, understanding, opinion, sense, from which comes every art and science, and we ourselves become reasoning beings."[21]

William Rowan Hamilton was fluent in Greek and Latin, and he had read many classic works. What had he read on Pythagorean lore? At least he had read works by the witty satirist Lucian, who in particular referred to the Tetractys or Quaternion, "which is their most solemn oath, and sums their perfect number, the name Beginning of Health."[22] In a fictional dialogue, Lucian made one character, Critias, critically refer to a "Tetractys" of Pythagoras: "I do not understand your one three, and three one; you might as well talk of the Tetractys of Pythagoras, his four, his eight, and his thirty."[23]

Hamilton shared his "Tetractys" sonnet with several friends. When he showed it to Professor William Archer Butler, of Dublin, "a poetical and philosophical friend," Butler replied: "I see clearly now that your Quaternions are a gross plagiarism from Pythagoras." By contrast, Hamilton considered it not a plagiarism, but instead an *"acknowledgement."*[24] Having enjoyed his classical education, Hamilton believed that the ancient Greek philosophers had achieved some of the world's most wonderful discoveries. His desire to believe shows up in another comment he made in 1852: "May we believe these stories? I hope we may at least believe that Pythagoras discovered the *property* of the hypotenuse—whatever becomes of the story of the *hecatomb*."[25]

In America, a commentator praised Hamilton for the lonely quest that had led him to discover an eternal truth: "It is the same true spirit of the geometer which led Pythagoras, twenty-three hundred years earlier, to offer a hecatomb in gratitude to the gods for the discovery of a single new proposition in regard to the right triangle. And if the world should stand for twenty-three hundred years longer, the name of Hamilton will be found, like that of Pythagoras, made immortal by its connection with the eternal truth first revealed to him." The commentator said that by contrast to poets, "Whatever the mathematician really imagines, is not imaginary, but real."[26]

Meanwhile, Hamilton worked to validate quaternions. One way to argue that his system was not an arbitrary invention was to show that it matched something in nature. Hamilton tried to analyze physical problems in terms of quaternions. The three imaginary terms could represent three-dimensional magnitudes, such as velocities. Hamilton called the three-part portion of a quaternion "a vector," and he referred to i, j, k as unit vectors. But if the one real term in a quaternion represented a time, as he originally suggested, then why was this time *added* to a three-dimensional line segment? Also, the imaginary terms led to ambiguities. For example, physicist James Clerk Maxwell complained that in quaternions, kinetic energy was always negative. Kinetic energy was defined as $k.E. = ½ mv^2$. Since velocity, v, is a directed magnitude, it would be an imaginary number, so its square would be negative. Hence, some critics disliked quaternions.

In 1894, a Scottish physicist and mathematician described how some individuals had reacted to Hamilton's work. He said that "a Scottish mathematician, on reading Hamilton's Quaternions, first formed the alternative conclusion that either he himself was a dull stupid or the book sheer nonsense, but on reading further was able to arrive at the more comforting alternative; that a German mathematician declared the method to be 'an aberration of the human intellect'; and that a French mathematician gave the verdict, 'Quaternions have no sense in them, and to try to find for them a geometrical interpretation is as if one were to turn out a well-rounded phrase, and were afterwards to bethink oneself about the meaning to be put into the words.'"[27]

Meanwhile, regarding the neglect of quaternions in physics, another writer complained: "It is a curious phenomenon in the History of Mathematics that the greatest work of the greatest Mathematician of the century which prides itself upon being the most enlightened the world has yet seen, has suffered the most chilling neglect."[28]

Nevertheless, some eccentric individuals tried to modify Hamilton's scheme, to adjust it to physics. Oliver Heaviside worked on this puzzle. Heaviside grew up in poverty in London, and illness impaired his hearing. His father regularly beat him, and his mother was a sour schoolteacher. He left school at the age of sixteen and worked as a telegraph operator for a few years but then quit at the age of twenty-four. He lived with his parents in London, unemployed, for fifteen years, until they moved with him to the small, coastal town of Paignton, in southwest England, in 1889. Heaviside was shy and withdrawn, with piercing eyes that frightened children. He had no college degree and was never affiliated with any university. Still, he gradually taught himself mathematics and physics, including Maxwell's theory of electricity and magnetism, which discussed quaternions. Heaviside wanted to "murder" some aspects of Maxwell's theory, and he also disliked some aspects of quaternions.[29] He joked that an American schoolgirl defined quaternions as "an ancient religious ceremony."[30]

Heaviside appreciated vectors, as Maxwell had used them, but he saw no need to justify the rules of vectors in terms of quaternions: "The laws of vector algebra themselves are established through Quaternions, assisted by the imaginary $\sqrt{-1}$. But I am not sure that any one has ever quite understood this establishment. . . . I never understood it."[31] In quaternions, the square of a unit vector was −1, which Heaviside called a "convention." He complained that "some of the properties of vectors professedly proved were wholly incomprehensible. How could the square of a vector be negative?"[32] The negative sign in any vector squared was "the root of the evil."[33]

Meanwhile in Connecticut, at Yale, professor Josiah Willard Gibbs independently made the same complaints. Gibbs had studied engineering but became a physicist and chemist. He too was a life-long bachelor, spending much time alone, though he lived with his sister and her family. He had bright blue eyes and was kind and unassuming, despite his many contributions to science. He was a patient and punctual man who fulfilled his duties conscientiously. At Yale, he was known as "the man who never made a mistake."[34] He studied mathematics, primarily for its practical uses in science. In 1877, Gibbs founded the Yale Mathematical Club, and at one of their meetings he commented: "A mathematician may say anything he pleases, but a physicist must be at least partially sane."[35]

Gibbs and Heaviside both thought that vectors were very useful in physics, but not quaternions. They decided to formulate an algebra of vectors, while excluding the confusing fourth term. They took vectors such as

$$ix + jy + kz$$

and decided to change the way in which the imaginary terms are squared. Traditionally, everyone had claimed that

$$i^2 = -1$$

But now, Gibbs and Heaviside decided that

$$i^2 = +1$$

and so too for j and k, in order to eliminate the physically meaningless negative sign that had bothered Maxwell. Gibbs and Heaviside realized that they could also delete all reference to imaginaries, that is, use i, j, k as unit vectors without saying that they are imaginary numbers.

Using his own money, Gibbs self-published a pamphlet on his "Vector Algebra," and Heaviside too published his approach. By changing the rules of i, j, k, Gibbs and Heaviside violated some of the algebraic elegance of Hamilton's original scheme. In particular, in Hamilton's system, the multiplication of quaternions obeyed the associative property. If we apply Hamilton's multiplication rules we get

$$i \times (j \times j) = (i \times j) \times j$$
$$i \times (-1) = k \times j$$
$$-i = -i$$

This associative law, $(ab)c = a(bc)$, remains valid for all kinds of quantities. By contrast, using the rules of Gibbs and Heaviside, the multiplication of i, j, k is not associative:

$$i \times (j \times j) \qquad (i \times j) \times j$$
$$i \times (0) \qquad (k) \times j$$
$$0 \neq -i$$

Consequently, some fans of quaternions shunned the new vector rules. One prominent critic ridiculed Gibbs's scheme as "a sort of hermaphrodite monster."[36] But Gibbs defended his system. And from the small town of Paignton, Heaviside commented: "There is confusion in the quaternionic citadel; alarms and excursions, and hurling of stones and pouring of boiling water upon the invading host."[37]

Meanwhile, Heaviside's simple life gradually fell apart. His mother died in 1894, and his father died two years later. In 1897 Heaviside moved to another house, but he soon became very ill. Neighborhood boys ridiculed him, yelling insults, and he complained that they spied on him. He asked policemen to help, but they did not. Heaviside became increasingly ill, but boys continued to harass him, throwing rocks at his house: "Panes broken and splashed over my sickbed."[38] Meanwhile, Gibbs had died in 1903, of a sudden intestinal ailment.

In time, other researchers developed other systems that also became known as vector algebras. Whereas Hamilton believed that he had discovered the one true system for the analysis of space, some other mathematicians and physicists devised new systems involving neither quaternions nor the restriction of dealing only with three-dimensional space.

The history of vector systems illuminates the growth of mathematics, partly because such developments are relatively recent and therefore well documented. Teachers often introduce negative or imaginary numbers as if such things had always existed, unknown, waiting to be recognized for ages. But the origins of quaternions, octonions, and the Gibbs-Heaviside algebra reveal a neglected topic: the design of concepts and rules. Repeatedly, there were clear-cut rules that seemed universally valid, for example, the laws of multiplication, until some eccentric individuals broke such rules and thus managed to create new kinds of mathematics.

Did Hamilton invent new imaginary numbers, or did he discover them? He certainly believed that he had discovered them, that these numbers had existed for thousands of years, in some sense waiting quietly for somebody to reveal them in nature. Accordingly, a memorial stone plaque on the stone wall on the east side of Broom Bridge (now *Broom* with no *e*) reads:

> Here as he walked by
> on the 16th of October 1843
> Sir William Rowan Hamilton
> in a flash of genius discovered
> the fundamental formula for
> quaternion multiplication
> $i^2 = j^2 = k^2 = ijk = -1$.
> & cut it on a stone on this bridge

But instead, I am convinced that Hamilton's imaginary numbers were a product of invention. If we take into account Hamilton's *j* and *k,* along with combinations of the several imaginary numbers described by Graves and Cayley, we can write multiple square roots of –1:

$$\sqrt{-1} = i, \quad \sqrt{-1} = j, \quad \sqrt{-1} = k$$

$$\sqrt{-1} = -i, \quad \sqrt{-1} = -j, \quad \sqrt{-1} = -k$$

$$\sqrt{-1} = l, \quad \sqrt{-1} = m, \quad \sqrt{-1} = n, \quad \sqrt{-1} = o$$

$$\sqrt{-1} = -l, \quad \sqrt{-1} = -m, \quad \sqrt{-1} = -n, \quad \sqrt{-1} = -o$$

$$\sqrt{-1} = \sqrt{ijk}, \quad \sqrt{-1} = \sqrt{ilm}, \quad \sqrt{-1} = \sqrt{ion}, \quad \sqrt{-1} = \sqrt{jln}$$

$$\sqrt{-1} = \sqrt{jmo}, \quad \sqrt{-1} = \sqrt{klo}, \quad \sqrt{-1} = \sqrt{knm}$$

Are there more square roots of –1? More stunningly, we should ask: what is it about –1? Why does *this particular number* have so many square roots?

The common habit is to ignore these questions, to disregard hypercomplex numbers, so that basic imaginary numbers convey a simpler impression, more consistent with real numbers. But these important questions should be raised. At bottom, the multiplicity of roots is a byproduct of the rule that minus times minus is plus. That one rule led to the notion that a single number can have two square roots and, more generally, that a *single* operation of root extraction can lead to *multiple* distinct results.

Nowadays, writers and teachers still mostly speak as if there is only one solution for the operation $\sqrt{-1}$, a number called *i*, "the square root of –1." This comfortable myth was criticized, for example, in 1884, when the mathematician and logician Gottlob Frege explained: "Nothing prevents us from using the concept 'square root of –1'; but we are not entitled to put the definite article in front of it without more ado and take the expression 'the square root of –1' as having sense."[39]

Regardless, many popular writers, such as Paul Nahin, claim that the square root of –1 really signifies perpendicularity and that "there is nothing at all imaginary about $\sqrt{-1}$."[40] This same opinion was voiced by Arnold Dresden, president of the Mathematical Association of America from 1933 to 1934, who also asserted the "reality" of imaginary numbers. Dresden proposed that imaginary numbers should instead be called "normal numbers," partly because *normal* is a synonym for perpendicular but also "in the hope that it will divest these perfectly innocent numbers of the awe-inspiring mysteriousness which has always clung to them."[41] In agreement, another mathematician said that the expression "normal numbers" conveys "a much healthier sound than 'imaginary.'"[42]

These people talk about imaginary numbers as if they are objects, not concepts. Such educators write as if there is something dreadful in the idea that some mathematical elements might be products of the imagination. But the historical origins of certain concepts show that imagination and invention played important roles in their development. Is there anything imaginary about imaginary numbers? Yes, and it's actually a *good* thing—there is nothing unhealthy about using imagination in mathematics.

Mathematicians eventually accepted Hamilton's imaginary numbers as logically consistent, but some viewed them with discomfort. Likewise, octonions have not received the appreciation that would lead many educators to teach them to students. In view of such reservations, the mathematical physicist John C. Baez wryly commented: "The real numbers are the dependable breadwinner of the family, the complete ordered field we all rely on. The complex numbers are a slightly flashier but still respectable younger brother: not ordered, but algebraically complete. The quaternions, being noncommutative, are the eccentric cousin who is shunned at important family gatherings. But the octonions are the crazy old uncle nobody lets out of the attic: they are *nonassociative*."[43]

Some writers pitied Hamilton for having spent the last twenty-two years of his life, from 1843 until he died in 1865, working almost exclusively on quaternions. For example, in 1937, Eric Temple Bell, also a former president of the Mathematical Association of America, remarked: "Hamilton's deepest tragedy was neither alcohol nor marriage but his obstinate belief that quaternions held the key to the mathematics of the physical universe. History has shown that Hamilton tragically deceived himself when he insisted ' . . . I still must assert that this discovery appears to me to be as important for the middle of the nineteenth century as the discovery of fluxions [the calculus] was for the close of the seventeenth.' Never was a great mathematician so hopelessly wrong."[44]

It's true that quaternions have not become as useful as calculus; however, they now are far more common than in 1937. Not only are quaternions used in various fields of physics, but quaternions are now appreciated as one of the most important concepts in computer graphics. They constitute a powerful way to represent and compute rotations in three-dimensional space. By comparison to rotation matrices, they require less memory, compose faster, and are well suited for efficient interpolation of rotations. Accordingly, aircraft and spaceships use quaternions to compute their motions and orientations in space. To represent the attitude of an object, mathematics provides various approaches: a sequence of rotations known as Euler angles, a single 3 × 3 matrix known as a direction cosine matrix, or a single axis and angle using Euler's theorem. But such methods often represent angles in terms of components; quaternions serve to represent and operate with angles and axis attitude directly. Quaternions offer some striking advantages over other approaches: quaternion operations involve no trigonometric functions; they are less susceptible to errors in rounding up numerical values; by consisting of four pieces of information, they are more compact than direction cosine matrices (which involve nine pieces of information); and quaternions vary continuously over the range of all possible attitudes (there are no quaternion singularities, as when dividing by zero). Performing mathematical operations with quaternions can be tedious, but such operations are easy to program, and computers are excellent for carrying them out and faster than when computing direction cosine matrices. For these reasons, quaternion mathematics is now *the* standard system used by computers operating the navigation system of aircraft. Whenever you fly in an airplane, its attitude computations are carried out by quaternions!

Quaternions are valuable in computer graphics, aircraft, *and spaceships,* so Hamilton was right: in the future, his system became very important. But also of great importance is the fact that their value lay beyond their scientific and technological applications; they showed that it is possible to develop new kinds of numbers and algebras that change some of the rules of traditional algebra. This finding was so compelling that now there exist many algebras where previously, for centuries, there had existed only one.

9

THE WAR OVER THE INFINITELY SMALL

Scientists used to say that matter is made of indivisible units, atoms. But some thought that matter is divisible into fragments much smaller. In 1896, physicist Emil Wiechert commented: "We might have to forever abandon the idea that by going toward the Small we shall eventually reach the ultimate foundations of the universe, and I believe we can do so comfortably. This universe is indeed 'infinite' in all directions, not only outward in its Greatness, but also down, into the Smallness within."[1] Soon, radioactivity seemed to show "the almost infinite divisibility of matter."[2]

While physicists divided nature into subatomic particles, mathematicians divided imperceptible quantities into bits unimaginably smaller. But they argued about these infinitesimal pieces, numbers that were apparently impossible, inconceivable, insane. They imagined numbers so small that a great sum of them made no difference whatever, numbers that resemble both infinity and zero, nearly as negligible as nothing.

Before discussing their debates, let's first consider similar disagreements among nonmathematicians. Each year at the University of Texas at Austin, I start one of my courses by asking students to answer a survey with several questions, such as:

According to mathematicians, which of these propositions is true?

☐ .999 . . . is equal to 1.

☐ .999 . . . approaches 1 but does not reach it, because 1 is its limit.

The ". . ." means that the numeral 9 continues to repeat infinitely. What do you think? Is .999 . . . *equal* to 1?

Most of my incoming students, mostly math majors, plus many upper-level science students, check the bottom option. They believe that .999 . . . is smaller than 1. Regardless of who's right or wrong, it's interesting that here students disagree about a basic question. So I ask them: why do you think that they're *different*? They answer:

"Because they're *obviously* different . . ."

"Because .999 . . . approaches 1 but never reaches it . . ."

"There's a physical difference between a whole and a bit less . . ."

"Because 0.999 . . . is a *fraction,* because there's nothing on the left of the decimal point."

"The two numbers are next to one another, but are not the same . . ."

When I asked about this last statement, some students said that there are infinitely many numbers between the .999. . . and 1. Also, some students have said, "If we subtract .999 . . . from 1, we get a tiny little bit left, so the two are not equal."

However, if you happen to believe that 1 = .999 . . . then you may feel that all these comments are mistaken. One might be sure enough to say: "*Belief has nothing to do with it—they are the same.*" And maybe your reasons for regarding the two as utterly *equal* resemble the reasons given by some students. For example, Jess, a math major, quickly wrote on the blackboard:

$$\frac{9}{9} = 1$$

$$9 \times \frac{1}{9} = 1$$

$$9 \times .111 . . . = 1$$

$$.999 . . . = 1$$

And then Jess smiled big for having apparently given an elegant proof of a basic truth; from 9 = 9 it follows that .999 . . . = 1.

But let's pause on one step. How do we know the following?

$$\frac{1}{9} = .111 . . .$$

We implicitly begin to divide as follows:

$$\begin{array}{r} .1 \\ 9\overline{)1.0000000} \\ \underline{-9} \\ 1 \end{array}$$

The division must continue indefinitely because there is a remainder of 1, which must be divided by 9. Since it seems that there is always a remainder that has not yet been divided, it is unintuitive to some students to imagine that any series such as .111111111 . . . , even if it is infinitely long, actually is *equal* to 1 divided by 9. Let me clarify this impression. Here are some statements on which everyone agrees:

We all agree that .1 *is not equal* to 1/9.

We all agree that .11111 *is not equal* to 1/9.

We all agree that .111111111111111111 *is not equal* to 1/9.

Thus, some students think that if we keep adding 1's we still do not have 1/9, even if we add infinitely many 1's. Meanwhile, other students accept that *if we have* infinitely many 1's then it *is* equal to 1/9.

Hence students disagree on their intuitions about what happens at infinity. Likewise, mathematicians have had nasty disagreements and arguments about infinity, for centuries. Nowadays, nearly all mathematicians do agree that .999 . . . = 1: that's the official answer. Explanations abound; some are more or less satisfying than others.

Here's a simple algebraic proof. Let x stand for .999 . . . , and let's find x:

$x = .999 . . .$

Multiply both sides by 10:

$10x = 9.999 . . .$

subtract the same from both sides:

$10x - a = 9.999 . . . - a$

and this same quantity a can well be

$10x - x = 9.999 . . . - .999 . . .$

which gives

$9x = 9$

$x = 1$

and therefore

$1 = .999 . . .$

Given this argument, more students feel compelled to agree. In 1770, a similar example was published by Leonhard Euler.[3] He argued:

> There is a great number of decimal fractions, therefore, in which one, two, or more figures constantly recur, and which continue thus to infinity. Such fractions are curious, and we shall show how their values may be easily found. Let us first suppose, that a single figure is constantly repeated, and let us represent it by a, so that $s = 0.aaaaaaa$. We have
>
> $$10\,s = a.aaaaaaaa$$
>
> and subtracting
>
> $$s = 0.aaaaaaaa$$
>
> we have
>
> $$9\,s = a; \text{ wherefore } s = a\,/\,9$$

Therefore, if we choose to set $a = 9$, we have again: $1 = 0.9999999\ldots$, while if we set $a = 8$, as another example, we have: $s = 8/9 = 0.8888888\ldots$, and by multiplying both sides by 9, we have: $8 = 7.9999999\ldots$.

Another argument proceeds as follows. This one uses no algebra, just numbers. Suppose we agree that $1/3 = .333\ldots$, and we agree that

$$\frac{1}{3} + \frac{1}{3} + \frac{1}{3} = \frac{3}{3} = 1$$

We should therefore agree that

$$.333\ldots + .333\ldots + .333\ldots = .999\ldots = 1$$

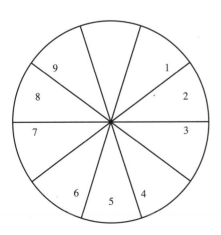

Figure 9.1. A pie cut in ten slices.

Still, someone might object to this kind of argument by doubting whether $1/3 = .333\ldots$. What would it mean, *physically*, to divide 1 among 3 and to get $.333\ldots$? Some of us imagine the following procedure. Take one thing, such as an apple pie, and cut it into ten slices. Now, distribute the slices among three persons by giving three slices to each, for a total of nine slices. But there is one slice left on the table.

So, now cut that one slice into ten small slices, and distribute them equally to each person, so each person gets three small pieces—but there is one small piece left.

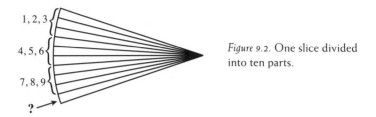

1, 2, 3

4, 5, 6

7, 8, 9

?

Figure 9.2. One slice divided into ten parts.

Okay, now cut that small piece into ten smaller pieces and distribute those too. There is still one of those pieces left on the table. Continue to repeat this procedure, even infinitely. Will there ever be nothing left on the table?

Some of us might start to wonder about the limits of *physical* division. We will reach a point when we are not dividing pie, just atoms, electrons, or other subatomic particles, and at some point that will become problematic. But the point of the illustration is just to exhibit the intuition that many of us have: that if we carry out divisions by ten, followed by divisions by three, we will *never* run out entirely of what we started to divide. There will always be something left, because of the very *procedure* for dividing. We are cutting in such a way that we will never finish cutting. Thus .333 . . . seems to represent this endless procedure.

To that concern, mathematicians might say that by writing 1/3, one is not cutting anything at all. They argue that .333 . . . is just the way of writing "one-third" in decimal notation. And .999 . . . is an alternative way of writing 1 in the decimal system.

To illustrate this, suppose that instead of using the base-10 number system we happen to use a base-60 system. In base 60, each unit is divided into sixty parts, instead of our usual ten. Thus, we might express one-sixtieth of one unit as .01. And we might write one-third of one unit as .20, such that

$$.20 + .20 + .20 = .60 = 1$$

Here we see that in this sexagesimal notation, one-third is not an infinitely repeating number; it's just a definite number, like 1/2 = .5 in the decimal system. So, if we accept that .333 . . . is just the effective if awkward way of expressing 1/3 in the decimal system, then likewise, .999 . . . can be accepted as a way of writing 1 in the same system.

By this point it might seem that the issue is settled. These last examples might effectively compel students into accepting that .999 . . . and 1 are equal. But actually, that's not my intention. To the contrary, now that we have some solid arguments for why .999 . . . = 1, and, now that we know on authority that *mathematicians agree* on this, let me doubt it by explaining why people who remain skeptical (nonmathematicians) are not being silly. I will give just two arguments, the second being the bigger one.

First, we have already mentioned one haunting intuitive argument, which we may rephrase as follows. Take the number .9, which we all acknowledge to be less than 1. Now add a small enough fraction, .09, so that the result continues to be less than 1. Now add a smaller fraction, .009, such that the result is still less than 1. And so on—repeat this process as much as you want. Clearly, the result will continue to be less than 1, just because the procedure by which we add new digits is defined in such a way that the result remains less than 1. People who doubt that .999 . . . = 1 construe ".999 . . ." as being the sum of a series of numbers that are *chosen to be less than 1.* This account substantiates a strong intuition at the root of the resistance. Arguments can be raised against it, as a mathematician might argue, for example, that .999 . . . is really not a construction of a series of digits added onto one another in time; it is just a symbol for a single number that is given all at once, like all other numbers. Some others argue that the expression ". . ." is not well-defined.

But consider now the second argument. Take the number 1, and subtract from it *an infinitely small quantity,* which we may write as *dt.* Now write, in decimal notation, the result. What will you write? We might be very tempted to write

$$1 - dt = .999 . . .$$

And here it is not easy to shrug off the idea that 1 ≠ .999 . . . . It now would seem that 1 is greater than .999 . . . because we have required that

$$1 = .999 . . . + dt$$

We reach the topic of the infinitely small. Here mathematicians fought over something very nearly close to nothing. Let's see how they have disagreed on whether the infinitely small is equal to zero. To do so, we'll consider debates in the history of the differential calculus.

Early on, most mathematicians agreed on the utility of the calculus. But many argued at length over the meaning of its basic rules. Consider an example. The following lines might seem obscure, but that's the point. You'll see how much mathematicians have argued over things that teachers try to teach as if they were clear as glass. So—we learn that there is something called a "derivative," and we learn rules for taking derivatives. For example, for the equation

$$s = 16t^2$$

we take its derivative by applying the "power rule":

$$\frac{ds}{dt} = n \cdot at^{n-1}$$

which gives

$$\frac{ds}{dt} = (2)16t^{2-1}, \quad \text{that is:} \quad \frac{ds}{dt} = 32t$$

Now, what do these expressions mean? How are they justified? Mathematicians have argued about this for hundreds of years.

To start, let's look at how such equations originally were connected to physical things. In the 1590s, Galileo studied how falling bodies speed up as they fall. Schoolbooks say that Galileo carried out experiments by dropping objects from the Leaning Tower of Pisa, but that's a myth.[4] It began in a confused note that Galileo's last secretary wrote in a manuscript. By the way, in accord with the Pythagorean idea of the transmigration of souls, his secretary also misrepresented the date of Galileo's birth, so that it would seem to follow the death of the great Michelangelo. In any case, owing partly to Galileo's various actual experiments, we know that a falling stone travels roughly

16 feet after 1 second,

64 feet after 2 seconds,

144 feet after 3 seconds, and so on.

Looking for a pattern in such numbers, Galileo realized that the distance the stone falls is 16 times the square of the number of seconds it falls:

$$16 = 16 \times 1^2$$
$$64 = 16 \times 2^2$$
$$144 = 16 \times 3^2$$

This recurring relation can be summarized by the equation:

$$s = 16t^2$$

where s stands for a distance and t stands for seconds. This is the equation we saw above; it appears in thousands of books. But looking at it, we may feel a bit uncomfortable: one side is a distance, the other a time. How can it be that by squaring a *time* and then multiplying it by 16 we get a *distance*? One way to make sense of it is to say that it establishes a *numerical* relation: it does not convert seconds into feet; it only compares their quantities.

Next, we saw from the distances covered in each second of free fall that the stone accelerates as it falls. The distance covered during the third second alone is

$$144 - 64 = 80 \text{ feet}$$

Hence the average speed during that second is

average speed = 80 feet/second

But physicists asked, what is the speed *at the end of that second?* That is, what is the speed of the stone right when three full seconds have passed?

Since we obtained the average speed during the third second by considering the positions of the stone at the beginning and the end of that one second, we might expect to get the speed at a single instant by taking the speed at only the final position. But since the concept of speed depends on a body traveling *some distance,* we can't just take a single position; there must be at least a very small distance traveled, so that we may divide it by something, even by an extremely small instant of time.

Therefore, mathematicians realized that they could imagine speed at an instant of time, the so-called instantaneous speed, as the ratio of an extremely small distance, *ds,* and an extremely small span of time, *dt:*

instantaneous speed = ds/dt

I say "extremely small," but we will soon see why most mathematicians said no such thing, why they instead said strange things such as "infinitely small."

So, to quantify the distance *ds,* we need its two endpoints. One of its endpoints is 144, since that is the location of the stone after the three seconds in question. And we can take the other endpoint to be an extremely small distance away from 144. Although we want to know the speed of the stone when it's located precisely at 144 feet, we're now considering also a position slightly beyond that, because we imagine that that second position is extremely close to 144. So, if we accept that assumption, given the equation

$$16t^2 = s$$

we know that the position of the stone at 3 seconds is

$$16(3)^2 = 144$$

And its position just a tiny fraction of time later, call that $3 + dt$ seconds, is

$$16(3 + dt)^2 = 144 + 96dt + 16dt^2$$

By subtracting the prior position, 144, we get the distance between them:

$$96dt + 16dt^2$$

That's the tiny distance traveled by the stone in the extremely small span of time dt. And since speed is distance divided by time, we may write

$$speed = \frac{96dt + 16dt^2}{dt}$$

which gives

$$speed = 96 + 16dt$$

That's the speed of the stone at about 3 seconds (that is, during the interval between 3 seconds and $3 + dt$ seconds). But it looks odd. We might still feel uncomfortable with the presence of t, a time, on the right side of the equation, while the left refers to a speed. But again, we can ignore that by remembering that we're dealing only with a purely *numerical* relation. Numerically, the speed is a bit greater than 96. So we might read it as saying that "the speed of the stone right after 3 seconds is just a tiny bit more than 96 feet per second."

But we began by asking: what is the speed of the stone at 3 seconds? Looking over the steps taken, we may reason that by making dt even smaller, we get closer to the speed of the stone at 3 seconds. Looking again at the expression

$$speed = 96 + 16dt$$

we readily see that if $dt = 0$, then the speed is 96. If we carry out the same procedure, and *if* we make $dt = 0$ at the very end, then we see that for 1, 2, and 3 seconds of falling time, the speed of the stone has the following values:

> after 1 second, it falls at 32 feet per second,
>
> after 2 seconds it falls at 64 feet per second,
>
> and after 3 seconds it falls at 96 feet per second.

There's a pattern in this sequence of numbers. By comparing them, these speeds, with the distance equation, $d = 16t^2$, mathematicians realized that the speeds may be calculated as follows:

$$32 = 2 \times 16(1)^{2-1}$$
$$64 = 2 \times 16(2)^{2-1}$$
$$96 = 2 \times 16(3)^{2-1}$$

And these equations can all be summarized by the equation

$$\frac{ds}{dt} = n \times a(t)^{n-1}$$

which is the "power rule" that we stated earlier. This rule generalizes a series of results. But remember that we are summarizing results such as

$$\frac{ds}{dt} = 96 + 16dt$$

which means that the power rule actually *omits* the puzzling term 16*dt*. Thus we reach the main subject of our discussion: how did mathematicians justify the omission of terms such as *dt*?

Let's see what the founders did. Consider Isaac Newton, who began to formulate the calculus in the 1660s. Again, hinting at the Pythagorean idea of transmigration, writers often say that Newton was born the year when Galileo died, but that's a mistake, which comes from using the Gregorian calendar to date Galileo's death, while using the Julian calendar for Newton's birth. Anyhow, how did Newton deal with the extra term *dt*? He proceeded as we did, first finding what we wrote as *ds/dt* (he wrote it as $\dot{x}$ and called it a "fluxion"), and he deleted the extra term, our 16*dt*, by saying that it is infinitely small and thus negligible. It was as if "infinitely small" were equal to zero, as if Newton were setting *dt* = 0, to discard it.

But there were problems with that. If *dt* = 0, then our initial expression 3 + *dt* is just equal to 3, and thus we do not have a distance, an interval between two positions, at all. Thus we cannot get a speed. Moreover, a ratio such as *ds/dt* seems impossible, because it then involves a division by zero.

Newton's views varied over the years, and he considered various ways to try to make sense of his mathematical procedures. Without going into details, we may note at least that Newton regarded an infinitely diminishing quantity as being greater than zero, so that it could function as a divisor yet be negligible when added. To avoid the ambiguities raised by such entities, Newton preferred to not let them stand in isolation, but to take them as ratios, because a ratio could be a finite quantity.[5]

Still, Newton continued to operate with "infinitely small" quantities and to delete them from results wherever he saw fit. Yet he increasingly tried not to use such infinitesimals. He knew that it was problematic to not uphold the utmost exactitude in mathematics; he commented: "errors are not to be disregarded in mathematics, no matter how small."[6]

Infinitesimals seemed to allow two interpretations: either they were greater than zero, or they were equal to zero. But both entail logical impossibilities:

if $dt > 0$ then it is false that $32 + 16dt = 32$,

and if $dt = 0$ then ratios such as ds/dt are impossible.

So how could Newton justify the claim, for example, that after 3 seconds the speed of a falling body is exactly 96 feet per second? To do so, he reasoned that this "ultimate velocity" happens neither before the body reaches its last location nor after, but at the moment it arrives.[7] It might seem that the instantaneous speed takes place at a single instant of time that is neither when $dt > 0$ nor when $dt = 0$, but in between the two, right when dt vanishes. Hence Newton referred to infinitesimals as "evanescent quantities."[8]

Impressed by Newton's works, one of his friends remarked: "Thales sacrificed an Ox for hitting on the method of inscribing a rectangled triangle within a circle. Pythagoras said he would give an Ecatomb for a trifling problem What then would he have given for Sir I. Ns inventions."[9]

Newton kept his calculus and his interpretations of it nearly secret, for decades. And when he finally began to publish them, in the 1700s, mathematicians celebrated his procedures, but many were puzzled and unsatisfied with his explanations. But earlier, while Newton secretly polished his version of the calculus, similar mathematical procedures were crafted independently in the 1670s by Gottfried Wilhelm Leibniz, in Germany. Leibniz was a prominent philosopher, lawyer, diplomat, and librarian who diligently taught himself mathematics. Leibniz was one of the few mathematical peers of Newton, so the two had several interests in common. For example, Leibniz too admired the ancients: "I have the greatest esteem for Pythagoras, and quite nearly, I believe that he was superior to all other ancient Philosophers, as he pretty much founded Mathematics and the Science of incorporeal things, having discovered that famous doctrine, an insight bright and worthy of a whole hecatomb, that all souls are permanent."[10] Leibniz too believed that the veiled knowledge of Pythagoras had been corrupted by incompetent interpreters. But Newton and Leibniz became bitter enemies as they anonymously accused each other of plagiarizing the calculus, wrongly, and each increasingly praised *himself*. Still, their views on the calculus were similar. They became bitter enemies partly because they both committed the same historical fallacy: *similar therefore same, same therefore borrowed.*

How did Leibniz deal with the extra terms, the infinitesimals? He supposed that infinitesimals were imaginary inventions, quantities smaller than any assignable number but greater than zero. Since these concepts gave results, he regarded them as "useful fictions."[11] To Leibniz, infinity and infinitesimals were not numbers.[12] But he expected that infinitesimals have "the same properties" as ordinary numbers. Yet he posited, for example, that

$$32 + 16dt = 32$$

He argued that quantities that differ only by an incomparably small quantity are *equal*.[13] Like Newton, Leibniz wavered on how to make sense of infinitesimals. He too emphasized the use of ratios rather than isolated infinitesimals. He too tried to replace infinitesimals with definite quantities. He too treated infinitesimals as distinct from zero but having properties of zero.

Again there was the problem of how one could begin by treating one quantity as if it were *greater* than zero and then treat it, in the same argument, as if it were *equal* to zero. Leibniz believed that everything in the world was constituted of "monads," the tiniest indivisible units of matter. The term had roots in the writings of Giordano Bruno, and tradition claimed that Pythagoras had spoken about the power of monads. Hence some writers have speculatively traced the roots of Leibniz's concept to that early source.[14]

Unlike Leibniz, some of his followers claimed that infinitesimals in fact *exist.* Johann Bernoulli claimed that infinitesimals are the components of incredibly tiny material particles, divided by God. Bernoulli believed that human reason discovers "pure" mathematical things, "which according to the healthy view of the Platonists are eternally in God and, just like Him, are not created."[15] And Bernoulli made perplexing claims: "A quantity diminished or enlarged by an infinitely smaller quantity is neither diminished nor enlarged."[16] His pupil Guillaume L'Hospital anonymously wrote the first textbook on the calculus, published in 1696, where he said that two quantities differing by an infinitesimal may be treated as equal.[17] L'Hospital regarded such quantities as simultaneously equal and unequal, and he too believed that infinitesimals were not an invention, but that they actually exist and had been *discovered.* When Leibniz voiced his disagreement, mathematicians begged him not to speak out, not to betray the cause.[18] But in letters he insisted: "I do not believe that there are or even that there could be infinitely small quantities, and that is what I believe to be able to prove."[19]

For centuries, mathematicians had rejected infinitesimals. We do not know who originated the notion of such ephemeral quantities. Historian Carl Boyer notes that the ancient Pythagorean discovery of irrationality (incommensurability)—that no multiple finite quantity will fit entirely in both the side and the diagonal of a square—might suggest an idea of some smaller kind of quantity, not finite, that might serve the purpose. Thus Boyer comes close to attributing the notion to the Pythagoreans, but gracefully he abstains: "We do not know definitely whether or not the Pythagoreans themselves invoked the infinitely small."[20] In any case, at around 375 BCE, Plato denied arbitrary divisibility: he claimed that there exist "wonderful numbers," units that are "equal, invariable, indivisible," and he reported that the masters of arithmetic "repel and ridicule anyone who attempts to divide absolute unity when he is calculating, and that if you divide, they multiply, taking care that one shall continue one and not become lost in fractions."[21] Later, *The Elements* seemed to deny the existence of infinitesimals.[22]

Likewise, at around 225 BCE, Archimedes worked mostly without using infinitesimals.[23] He expected that every number, however small, has the property that if it is added to itself many times, the result can be greater than 1. For example, 0.3 is smaller than 1, but if we add 0.3 to itself several times the result is greater than 1. Thus, for any given number *n,* Archimedes expected that *some* of the statements in the following series are true:

$$n > 1$$
$$n + n > 1$$
$$n + n + n > 1$$
$$etc. \ldots > 1$$

By contrast, for infinitesimals *all* of these statements would be false. Infinitesimals would be such that no matter how many times one adds them to themselves, the result would never be greater than 1.

Thus the notion of an infinitesimal seemed ridiculous, and mathematicians such as Newton and Leibniz avoided it. But by the 1690s, some mathematicians such as L'Hospital accepted infinitesimals as just another kind of number. But not everyone was pleased with the new outlook. One man who criticized infinitesimals in the calculus was George Berkeley, a bishop of the Church of England, at Cloyne, Ireland, since the 1730s.

Berkeley was annoyed to see that some mathematicians were losing faith in the Bible, as if some of the doctrines of Christianity were nonsense. In 1734, Berkeley published his booklet *The Analyst, A Discourse Addressed to the Infidel Mathematician.* There was a strange irony: some philosophers and mathematicians were skeptical of the Bible, claiming that they could not accept on faith statements that were not justified by reason, whereas, Berkeley realized, they did accept certain bizarre mathematical claims on the basis of the authority of Newton and Leibniz, as if by faith and against reason.

To Berkeley, faith in religion was appropriate, but in mathematics, it seemed repulsive and reprehensible.

Berkeley denied the existence of quantities infinitely small and equable to nothing. He insisted that either dt is equal to zero or it isn't. If $dt = 0$, then ds/dt is meaningless; if it isn't equal to zero, then $32 \neq 32 + dt$, and so 32 is *not* the speed. Berkeley complained that Newton's approach was problematic because it changed the meaning of terms midway through the argument. He complained that Newton contradicted himself by first supposing that a given quantity increases by a very small amount and afterward by supposing that the increment is zero.

Berkeley insisted that "the minutest Errors are not to be neglected in Mathematics. . . . Geometry requires nothing should be neglected or rejected."[24] And for him, mathematics should be a science concerned with things that we perceive. He objected to various annoying concepts in calculus: quantities that are infinitely smaller than any perceptible quantity, the division of things that have no magnitude, the notion of a velocity where there is no motion, the idea that a triangle can be formed in a point, and more.

Berkeley further argued that Newton's method was "obscure," "repugnant," and "precarious." Berkeley scorned mathematicians who accepted infinitesimals. He denounced the ratios that Leibniz wrote as ds/dt, Newton's so-called fluxions: "And what are these Fluxions? The Velocities of evanescent Increments? And what are these same evanescent Increments? They are neither finite Quantities, nor Quantities infinitely small, nor yet nothing. May we not call them the Ghosts of departed Quantities?"[25] Berkeley feared that it being easy to manipulate symbols, some mathematicians deceived themselves with expressions that implied contradictions or impossibilities or were empty of meaning. He rejected the expectation that anyone should submit to the authority of Newton and Leibniz, and he denounced it as a kind of idolatry and bigotry.

In place of faith, Berkeley called for critical thinking in mathematics. In his *Defence of Free-Thinking in Mathematics,* in 1735, he said:

> In my opinion the greatest men have their Prejudices. Men learn the elements of Science from others: And every learner hath a deference more or less to authority, especially the young learners, few of that kind caring to dwell long upon Principles, but inclining rather to

take them upon trust: And things early admitted by repetition become familiar: And this familiarity at length passeth for Evidence. Now to me it seems, there are certain points tacitly admitted by Mathematicians, which are neither evident nor true. And such points or principles ever mixing with their reasonings do lead them into paradoxes and perplexities.[26]

Berkeley believed that mathematics should be based on evident truths. He demanded that the principles of the calculus should be clear.

The submission to authority that Berkeley criticized was old: it had many earlier manifestations, for example, among the cult of Pythagoras. In 45 BCE, Cicero complained that teachers' authority tends to cripple students: "Indeed, often students eager to learn suffer the authority of their professed teachers as an obstacle, ceasing to apply their own judgment as they trust the judgment of their master. I generally reject the way of the Pythagoreans, who, when they affirmed some position in debate but were asked *why,* usually replied 'because He said it,' he being Pythagoras; so much did prejudiced opinion dominate, that authority prevailed unsupported by reason."[27]

Some mathematicians and philosophers realized that Berkeley's criticisms were incisive, even appropriate. Some mathematicians felt uncomfortable with the foundations of the calculus. Some decided to banish infinitesimals. They wanted to return to the kind of deductive rigor, from evident principles, that characterized Euclid's geometry. Alternative views had been sketched even by Newton and Leibniz. Newton suggested that instead of considering ratios of infinitely small quantities, one may consider a series of ratios of finite quantities that approaches a given limit.[28] Several mathematicians picked up on this vaguely formulated notion and tried to elucidate it. Their creative works were partly motivated by their teaching jobs, which encouraged them to try to present the elements of the calculus in a logical, intelligible format.

By the 1820s, Augustin Louis Cauchy, in Paris, managed to reformulate the calculus on the basis of the notion of limits.[29] Many students despised him as a notoriously awful professor, but at least he did struggle to clarify the foundations of the calculus.

Cauchy believed that mathematical concepts exist only as abstractions of physical things and have no separate existence.[30] He believed that matter is not infinitely divisible, so he rejected the notion of infinitely small quantities. He also believed that "infinity, eternity, are divine attributes that belong only to the Creator and that God himself cannot communicate to his creatures, not that his power is limited in any way, but because there would be a contradiction in terms, if the idea of infinity were applied to that which is susceptible to variation and change."[31] For Cauchy, there were no infinite numbers, no infinite lines, no infinitesimals.

Therefore, Cauchy defined a limit as the fixed numerical value that is approached indefinitely by a series of successive values. For example, let's express the ratio between intervals of space and time as

$$\frac{x_2 - x_1}{t_2 - t_1} = \frac{\Delta x}{\Delta t}$$

Since a body moves the total distance Δx in the time Δt, it moves a smaller distance in less time. We may consider a series of intervals of time, each smaller than Δt and each next interval smaller than the prior, and we can represent the distance covered in each time interval with some Δx. Thus we have a series of speeds,

$$\frac{\Delta x_1}{\Delta t_1} = v_1, \ \frac{\Delta x_2}{\Delta t_2} = v_2, \ \frac{\Delta x_3}{\Delta t_3} = v_3, \ \ldots \ \frac{\Delta x_n}{\Delta t_n} = v_n,$$

and we can continue the series until the intervals are as small as we want. The resulting sequence of speeds may be such that the variable v_n approaches a fixed numerical value L, so that the difference $|L - v_n|$ is as small as we want. If so, then L is said to be the limit of the sequence.[32] For Cauchy, the series could be imagined to extend "infinitely," but for him this meant only an indefinitely large series of values that become greater than any given number. And he defined infinitesimals not as fixed numbers but as *variables* that decrease indefinitely, approaching zero.

In this context, then, what is a speed? Even though Cauchy called the speed "the derivative," he did not view it as a quantity that was derived from the other variables (intervals of space and time). Accordingly, mathematicians defined instantaneous speed as the limit of the series above:

$$instantaneous \ speed = v = \frac{dx}{dt} = \lim. \frac{\Delta x}{\Delta t}$$

By choosing a Δt that is sufficiently small, we can make $\Delta x/\Delta t$ be as close to 96 as we wish. Here, the instantaneous speed v is "the last ratio of infinitely small increments." And that limit is now the fundamental concept; it is *by definition* that at 1 second the speed is exactly 32 feet per second. Thus, in this account

$$\frac{dx}{dt} \neq 96 + 16dt$$

By 1872, Karl Weierstrass refined Cauchy's theory of limits. Weierstrass was a mathematician at the University of Berlin, who had suffered for years from attacks of dizziness, sickness, and chest pains, apparently caused by the intensity of the efforts and time he spent on mathematics and teaching. Unsatisfied with Cauchy's theory, Weierstrass wanted to fully dispense with infinitesimals. He wanted to set the calculus on an entirely numerical basis, that is, by eliminating all references to geometrical reasoning. Furthermore, he wanted to eliminate all notions of motion and time from the foundations of the calculus, by trashing expressions such as "approaching the limit." For Weierstrass, x did not represent a sequence of successively changing values; instead, x stood for just a static set of numbers.

For Weierstrass, the "instantaneous speed" was the limit of a set of numbers. Again, this speed has not been calculated; it is not even a ratio, just a number distinct from a set of ratios. Following Weierstrass, one could still write, for the speed of our falling stone at 3 seconds:

$$instantaneous\ speed = v = \frac{dx}{dt} = L = 96$$

but here the term dx/dt is no longer to be understood as consisting of infinitesimals nor even as being a ratio at all, but just *an old-fashioned notation* for the limit.

Weierstrass succeeded in using only finite numbers, rather than infinitesimals, as well as in not taking any ratio $\Delta s/\Delta t$ where Δt is zero. However, the consequence of eliminating ideas of motion, space, and time from the definition of the derivative led to an unintuitive definition of speed:

> The instantaneous speed is v if the absolute value of $\Delta s/\Delta t - v$
> is less than any positive number p for all absolute values of Δt
> less than some other positive number p_2 (which depends on p
> and t).

Does that sound confusing? While banishing ideas about infinitesimals and convergence to zero, the calculus still seemed physically subtle or confusing. But at least it had become logically elucidated. The ancient Pythagorean aspiration to understand everything in terms of numbers seemed to have been fulfilled for the calculus.

Still, a few individuals continued to theorize about infinitesimals. And some leading mathematicians became increasingly annoyed. For example, Georg Cantor denounced infinitesimals as "cholera-bacillus" infecting mathematics.[33] He vigorously opposed infinitesimals as "absurdities" that were "self-contradictory and completely without use or benefit."[34] So he attacked the efforts of mathematicians who tried to justify them. Cantor ridiculed infinitesimals as impossible: "square circles," a kind of "sign-isticism," a play on symbols.

Looking back on the work of Karl Weierstrass, the historian of mathematics Carl Boyer, writing in the 1930s and 1940s, viewed the limit definition of derivatives as "the final definition," "the final elaboration."[35] He did not expect that the calculus had reached the ultimate development of its concepts, that it would cease to evolve. But it seemed that finally the calculus had been essentially freed of its roots in physical experience (notions of motion, time, continuity, change) and metaphysics (notions such as infinitesimals). Boyer claimed that infinitesimals, in particular, lacked logical justification and were ultimately unnecessary. Many mathematicians and philosophers agreed.

A myth developed to the effect that infinitesimals had been purged entirely from mathematics, at least until the mid-1900s. This story spread, for example, in a popular book by Philip Davis and Reuben Hersh.[36] I too used to tell this to my students. But as historian Philip Ehrlich has shown, some mathematicians working on geometry and functions, at least, did continue to develop theories about infinitesimal lines and numbers.[37]

Moreover, in the early 1960s, Abraham Robinson brought infinitesimals back into calculus. Robinson grew up in Germany but fled with his parents in 1933, when Hitler required that Jews be fired from their jobs as teachers and civil servants. Robinson later became a professor of mathematics at universities in Toronto, Jerusalem, and Los Angeles. In 1961, he introduced a new approach to the calculus that he called nonstandard analysis.[38]

Robinson knew that from the perspective of mathematical logic the system of real numbers was "incomplete," that is, that its fundamental rules admitted the existence of not only real numbers (positives, negatives, etc.) but also of strange "objects" that had not been contemplated in the usual accounts of mathematics. By logically reformulating the system of real numbers and axioms in a way that explicitly includes such strange objects, Robinson formulated an extended system.

Robinson posited the existence of "infinitely large numbers." Any such number is greater than any positive number. If p is any positive number, however large, then there are infinitely large numbers, such as n, such that for every case, $n > p$. By taking any such infinitely large number n, and placing it as the denominator in the fraction $1/n$, Robinson argued, the result is an infinitesimal, a number that is smaller than any positive fraction $1/p$ but is yet greater than zero. He logically demonstrated the "existence" of infinitesimals, actual numbers (not variables) greater than zero but lesser than any $1/p$. In his system, the statement that such strange numbers are not positive was inexpressible, and thus his argument avoided being obviously paradoxical. Robinson mistakenly believed that Leibniz had operated with infinitesimals. Still, more than two centuries after Leibniz's death, Robinson showed that infinitesimals indeed obeyed the rules of real numbers.

Once this expanded "numerical universe" had been formulated, infinitesimals became again valid mathematical entities. Thus an expression such as

$$\frac{dx}{dt} = 96 + 16dt$$

acquired a new and definite meaning in nonstandard analysis, namely that *the ratio* of two infinitesimal numbers is equal to the sum of a standard real number and a nonstandard number (an infinitesimal). Here the instantaneous speed is *not* the derivative:

$$instantaneous\ speed = v \neq \frac{dx}{dt} = 96 + 16dt$$

Instead, for Robinson, the instantaneous speed is just 96, "the standard part" of the ratio dx/dt. Henceforth, old but seemingly unjustifiable arguments in the calculus (such as convenient proofs that treated derivatives as ratios) became legitimate ways of doing mathematics again.

Summing up, mathematicians variously disagreed about infinitesimals and about how to make sense of the concept of instantaneous speed. The calculus arose from puzzles about space, time, and motion, that is, from reflections about physical experience. Yet following Plato, later mathematicians often tried to eliminate such fruitful and intuitive notions from calculus, to make it "independent" and "pure." Yet other mathematicians who analyzed physics and fictions were sometimes able to develop new and useful mathematics. Table 10 shows a selection of disagreements over the notion of infinitesimals.[39] We have seen that whether individuals regarded such notions as worthless fictions or as actual realities was not merely a philosophical issue; it sometimes led to distinct kinds of activity in mathematics.

Although mathematicians bickered over infinitesimals in the calculus, some of their disagreements were not pointless or sterile. Instead, they were *productive disagreements.* Some of the individuals who felt annoyed and unsatisfied by contemporary explanations were hence driven to devise their own

Table 10. A Selection, across the Centuries, of Notions about Magnitudes or Numbers Characterized as Infinitely Small

ca. 225 BCE	Archimedes	The multitude of lines in a figure is infinite.
Late 1600s	Isaac Newton	Infinitesimals are evanescent quantities.
1670s	Gottfried Leibniz	Infinitesimals are not numbers, they are useful fictions.
1696	Guillaume L'Hospital	Infinitesimals are numbers that actually exist; they were not invented.
1734	George Berkeley	Infinitesimals are repugnant fictions that should be eliminated from mathematics.
1755	Leonhard Euler	Infintesimals exist algebraically and are equal to zero. Calculus is the theory of reckoning with zeros; $dx/dt = 0/0$.
1820s	Augustin Cauchy	Infinitesimals do not exist; infinity is a property of God alone, but there are finite variable quantities that tend to zero.
1870s	Paul du Bois-Reymond	Infinitesimals exist, by logical necessity, as one of many orders of infinity.
1872	Karl Weierstrass	Infinitesimals and all notions of motion should be eliminated from the calculus.
1885	Wilhelm Killing	Infinitesimal line segments are impossible.
1889–90s	Giuseppe Veronese	Infinitesimals are relative quantities.
1870s–90s	Georg Cantor	Infinitesimals do not exist; they are impossible and are a horrible disease in mathematics.
1891	Giulio Vivanti	Infinitesimal line segments do not exist; they are unnecessary in the calculus, but similar entities can be defined by convention.
1892	Giuseppe Peano	Infinitesimal constant line segments are self-contradictory.
1899	David Hilbert	Infinitesimals are standard elements in mathematics.
1901	Bertrand Russell	Infinitesimals were rightly banished from mathematics.
1960s	Abraham Robinson	Infinitesimals logically do exist in the numerical universe and can be used in the calculus.

Note that these are summary statements, paraphrased, not necessarily quotations.

innovative approach for making sense of the formal procedures. Time and again there were those who felt that finally the calculus had matured to its ultimate form, that finally it was justified by logical foundations. Yet there remained others who disagreed, and to them we owe the later creative and constructive developments.

Abraham Robinson acknowledged that there is truth to the view that mathematics involves symbolic rules that are established at will and elucidated logically to their apparent consequences. So he argued: "I cannot imagine that I shall ever return to the creed of the true Platonist, who sees the world of the actual infinite spread out before him and believes that he can comprehend the incomprehensible."[40] Likewise, taking a long glance at history, Bertrand Russell complained of how often people have traditionally misconstrued mathematical knowledge:

> Most sciences, at their inception, have been connected with some form of false belief, which gave them a fictitious value. Astronomy was connected with astrology, chemistry with alchemy. Mathematics was associated with a more refined type of error. Mathematical knowledge appeared to be certain, exact and applicable to the real world; moreover it was obtained by mere thinking, without the need for observation. Consequently, it was thought to supply an ideal, from which every-day empirical knowledge fell short. It was supposed, on the basis of mathematics, that thought is superior to sense, intuition to observation. If the world of sense does not fit mathematics, so much the worse for the world of sense. In various ways, methods of approaching nearer to the mathematician's ideal were sought, and the resulting suggestions were the source of much that was mistaken in metaphysics and theory of knowledge. This form of philosophy begins with Pythagoras.[41]

After Pythagoras had been so often hailed as a hero in mathematics, finally a philosopher portrayed him as a scapegoat, the alleged culprit at the roots of Platonism. In his youth, Russell had believed that mathematical objects exist in a timeless way, independent of our minds, and are changeless, but he subsequently struggled for decades to break free of this "mysticism," a difficult but sobering process that he described as "a gradual retreat from Pythagoras."[42]

Returning to the issue of infinitely recurring decimals, it well seems that Robinson's infinitesimals resemble the sort of thing that students groping for a difference between 1 and .999 . . . try to articulate. Still, it remains standard practice to regard such expressions as equal. Mathematician Timothy Gowers, at Cambridge University, explains:

> We must once again set aside any Platonic instincts. It is an accepted truth of mathematics that one point nine recurring equals two, but this truth is not discovered by some process of metaphysical reasoning. Rather, it is a *convention.* However, it is by no means an arbitrary convention, because not adopting it forces one either to invent strange new objects or to abandon some of the familiar rules of arithmetic. For example, if you hold that 1.999999 . . . does not equal 2, then what is 2 − 1.999999 . . . ? If it is zero, then you have abandoned the useful rule that x must equal y whenever $x − y = 0$. If it is not zero, then it does not have a conventional decimal expansion (otherwise, subtract it from two and you will not get one point nine recurring but something smaller) so you are forced to invent a new object such as "nought followed by a point, then infinitely many noughts, and *then* a one." If you do this then your difficulties are only just beginning.[43]

Yet Gowers notes that the tricky difficulties are not insurmountable, thanks to Robinson's work. Gowers sees the equality as a convention, a definition, a stipulation: "1.9999999 . . . is the same number as 2. (About this last example, by the way, there can be no argument, since I am giving a *definition.* I can do this in whatever way I please, and it pleases me to stipulate that 1.999999 . . . = 2 and to make similar stipulations whenever I have an infinite string of nines.)"[44]

So again, is it true that .999 . . . = 1? Thinking about a single negligible infinitesimal, you might wonder why we write it as a couple of letters, such as dt. Why don't we write it as a decimal number? Or what if we subtract it from 1; what decimal number is that? We might be tempted to write: 0.99999 . . . . But since mathematicians decided that .999 . . . = 1, then that numeral is already taken. Thus we have no decimal numeral name for the infinitesimal subtraction. By including infinitely many numbers, mathematics seems to run out of decimals to express such numbers. Strange: one would have thought that there are infinitely many decimals to go around.

In the end, it's useful not to hide the seams in the elements of mathematics, but better to pull them out into the light, to appreciate them, and even to be annoyed by them. Something good might come out of it.

10

IMPOSSIBLE TRIANGLES

W hen Albert Einstein was a solitary boy, less than twelve years old, his uncle told him about the Pythagorean theorem. The boy struggled to confirm it until he devised a way to prove it to himself.[1] By reflex, one might be tempted to construe this anecdote as early evidence that Einstein was a genius, but no—he didn't see it that way, and there were already very many proofs of the hypotenuse theorem, made by ordinary people, young and old. What matters is that, as the young Einstein realized, certain geometric propositions seem compellingly true.

For example: "The sum of the angles of a triangle makes 180 degrees." That's something we learn in school, and it seems simple and true. Try to draw a triangle where the sum of the angles is *more* than 180 degrees, and we might get something like this:

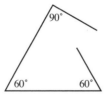

Figure 10.1. Is this a triangle?

But that's not a triangle. Instead, if you could draw a real triangle and chop it into three angles and add them up, they would make two right angles, 180 degrees, right?

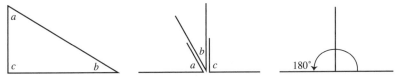

Figure 10.2. Given a triangle, the sum of its angles makes two right angles.

Since ancient times, geometers expected that this would be true for every triangle. But later, some mathematicians wondered: what if it *isn't* true? And what if it isn't true for *any* triangle?

But wait, geometers were sure that it's true because *they had proved it.* And we still prove it in school. For example, take any triangle whatsoever, and at one corner draw a straight line parallel to the opposite side, as illustrated.

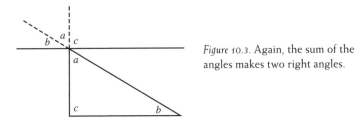

Figure 10.3. Again, the sum of the angles makes two right angles.

It's easy to grasp that the angles of the triangle are equal to the angles on the parallel line, as labeled, and therefore we conclude that the sum $a + b + c$ does make a straight line (two right angles). This construction, using parallel lines, can be carried out for any triangle, so geometers concluded that the sum of the angles of *any* triangle gives exactly the same result.

Such conclusions seemed so certain that people thought that geometry provided *universal* and *undeniable* truths. They combined religion with mathematics. In his *Republic,* Plato argued that "this knowledge at which geometry aims is of the eternal, and not of the perishing and transient. . . . Geometry will draw the soul towards truth, and create the spirit of philosophy, and raise up that which is now unhappily allowed to fall down."[2] At around 450 CE, the pagan theologian Proclus claimed that the ancient Pythagoreans had insinuated the transcendent nature of geometry: "They implied that the geometry which is deserving of study is that which, at each new theorem, sets up a platform to ascend by, and lifts the soul on high instead of allowing it to go down among sensible objects and so become subservient to the common needs of this mortal life."[3] Proclus also believed that the dialogues of Plato had been divinely inspired.

Meanwhile, Hippolytus complained that the Pythagoreans worshipped a Creator of all who allegedly was "the Great Geometrician and Calculator."[4] In the 1620s, Galileo argued that the universe is intelligible precisely because it is structured by geometrical figures. Kepler claimed that geometry is coextensive with God: "Geometry, which before the origin of things was coeternal with the divine mind and is God himself (for what could there be in God which would not be God himself?), supplied God with patterns for the creation of the world, and passed over to Man along with the image of God; and was not in fact taken in through the eyes."[5] In the 1640s, René Descartes too argued that mathematical truths did not originate in perceptions, but were truths that God had implanted in our minds since birth. In 1783, the philosopher Immanuel Kant claimed that geometry consists of propositions that are "thoroughly recognized as absolutely certain." Geometrical order was a necessary precondition for human thought, part of the structure of our mind, independent of our physical experience and perceptions. Triangles and parallel lines were part of the permanent architecture of our logical minds.

Figure 10.4. Is God a geometer?

The ancient arguments were based on claims that could hardly be doubted. For example, Euclid, or the authors of the *Elements,* listed several "Common Notions," such as, "Things that are equal to the same thing are also equal to each other," and "If equals are added to equals, then the wholes are equal." Philosophers used to think that *everyone* must agree to that. Such common notions were supposed to apply not only to geometry but to other fields of knowledge too.

Now, among purely *geometrical* claims, the *Elements* required many statements, including the following:

1. A straight line can be drawn between any two points.

2. Any straight line can be extended.

3. A circle can be drawn having any center and radius.

4. All right angles are equal.

These statements seem clear. The fourth says, for example, that the two right angles pictured are equal:

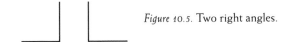

Figure 10.5. Two right angles.

The orientations are different, but the angles are equal. It might seem odd that such statements were described as "postulates," meaning assumptions that are not obviously true but are posited to be true. Yet the next postulate, the fifth, seems strange compared to the others:

5. If two straight lines on a plane are crossed by a straight line, making the interior angles on the same side less than two right angles, then, the two straight lines, extended sufficiently, meet on that side on which are the angles less than the two right angles.

This postulate seems convoluted, long, not as simple as the rest. It can be understood more quickly with a drawing, as shown in figure 10.6.

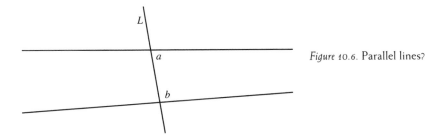

Figure 10.6. Parallel lines?

Two straight lines are crossed by line *L,* and *if* the internal angles on one side, *a* and *b,* sum to less than two right angles, then the two lines, if extended, must cross somewhere on that side.

The construction implies that for other lines, not pictured, *a* and *b* can be of such sizes that the two lines do not cross, no matter how much they are extended. Those would be called "parallel lines," and so the fifth became known as the parallel postulate. It was widely used, even to prove basic propositions such as the so-called Pythagorean theorem. In the 1830s, William Rowan Hamilton commented: "No candid and intelligent person can doubt the truth of the chief properties of *Parallel Lines,* as set forth by EUCLID in his Elements, two thousand years ago."[6]

But a few people did not like the parallel postulate. Once in a while, a geometer tried to derive it from the other four. But such geometers repeatedly failed at deriving the fifth; they failed to show that it was a theorem rather than a postulate. For example, in 1733, the Italian Jesuit Girolamo Saccheri argued that although "nobody doubts the truth" of the fifth, it was not self-evident; it needed proof. So he tried to prove it by showing that contrary claims lead to contradictions. But his arguments were inconclusive. By 1759, Jean D'Alembert referred to the many failures to prove the fifth postulate as "the scandal in the elements of Geometry."[7]

Meanwhile, some geometers tried to reword the postulate, to make it shorter. In 1785, for example, William Ludlam proposed the postulate: "Two straight lines which cut one another cannot both be parallel to the same straight line."[8] In 1795, John Playfair postulated: "Two straight lines cannot be drawn through the same point, parallel to the same straight line, without coinciding with one another," later crediting Ludlam.[9] Subsequently, other writers paraphrased these postulates as: "Given a straight line and a separate point, through that point there exists *only one* straight line parallel to the first straight line." And this sort of statement became known as "Playfair's axiom."

In 1799, at Göttingen, the young student Carl Gauss tried to prove the fifth postulate, but he began to doubt the truth of geometry. One of his friends, Farkas Bolyai from Transylvania, also tried to prove the parallel postulate. In 1804 he sent Gauss the product of his labors, but Gauss saw that it was flawed. Long struggles made Farkas Bolyai increasingly miserable and exhausted. In 1813 Gauss commented: "In the theory of parallel lines we are still not further along than Euclid. This is the shameful part of mathematics, which sooner of later must take an entirely different shape."[10] Consequently, Gauss concluded that the fifth postulate was independent from the others, and he secretly explored the consequences of discarding the fifth. Gauss imagined an "anti-Euclidean geometry." Meanwhile, he became a well-known mathematician, but he rarely voiced skeptical comments about the truth of traditional geometry. He did not publish anything about his anti-Euclidean geometry. He feared that geometers would ridicule such ideas, with the "shrieks of the Boetians," an ancient tribe whose members, to the Greeks, seemed to be idiots.[11]

Yet several other eccentrics also pondered the consequences of denying the fifth. One, a professor of law, Ferdinand Karl Schweikart, outlined a scheme, free of Euclid's "hypothesis," in which the angles of a triangle would make less than two right angles. In 1818 Gauss received Schweikart's brief outline of "Astral Geometry," and he sympathized.[12] Meanwhile, in Transylvania, the son of Farkas Bolyai struggled along his father's path.

Farkas had taught mathematics to his son János, who learned it quickly and eagerly, in his father's words, "like a demon."[13] As a thirteen-year-old, János Bolyai sometimes taught his father's classes. His mother was often ill: she became insane for four years and died. János became a student of military engineering in Vienna; he excelled at sword fighting, the violin, and mathematics.[14] His father increasingly worried that his son too would waste his time and health on the unsolvable parallels. After some failures, by 1820, the son conjectured that *maybe* the parallel postulate could not be proven because it was just *false*. His father became horrified:

> You must not attempt this approach to the parallels: I know this
> way to its very end. I have traversed this bottomless night, which
> extinguished all light and joy from my life. For God's sake! I beg
> you to leave the parallels alone, abhor them like indecent talk, they
> may deprive you from your time, health, tranquility and the happi-
> ness of your life. That bottomless darkness may devour a thousand
> tall towers of Newton and it will never brighten up the Earth. . . . I
> thought I would sacrifice myself for the sake of the truth. I was ready
> to become a martyr who would remove the flaw from geometry and
> return it purified to mankind. I accomplished monstrous, enormous
> labors; my creations are far better than those of others yet I have not
> achieved complete satisfaction. . . . I turned back when I saw that
> no man can reach the end of this night. I turned back distraught,
> pitying myself and all mankind. . . . I have traveled past all reefs of
> this infernal Dead Sea and have always come back with broken mast
> and torn sail. The ruin of my disposition and my fall date back to
> this time.[15]

Yet János Bolyai obsessively struggled, and in 1823 he wrote to his father: "I have discovered things so wonderful that I was astonished, and it would mean everlasting shame to let them be lost forever; my dear Father, if you see them you will acknowledge them; but now I cannot say anything else: out of nothing I have created a strange new world; everything else that I have sent you is just a house of cards compared to a tower."[16]

Likewise, another eccentric who explored the weird idea of changing ge-ometry was Nicolai Ivanovich Lobachevsky, in Russia. He reached essentially the same results as Gauss and Bolyai. There were several creative individuals, but let's look at the work of just one of them, because he wrote most clearly.

Lobachevsky studied at the University of Kazan, where he was a disrup-tive student who despised taking orders. He barely avoided expulsion and eventually managed to become a lecturer at the university. By 1815 he had begun working on the puzzle of whether Euclid's fifth was provable. For centuries, geometry had remained remarkably constant, and so, most mathematicians valued it as virtually perfect. Lobachevsky, however, interpreted the constancy of geometry as a kind of stagnation, as if geometry had made "no advance from the state in which it has come to us from Euclid." Lobachevsky claimed that the "imperfections" and "obscurity" in the elements of geometry were due partly to the traditional concept of parallel lines. So, he argued: "The fruitlessness of the attempts made, since Euclid's time, for the span of 2000 years, aroused in me the suspicion that the truth, which it was desired to prove, was not contained in the data themselves; that to establish it the aid of experiment would be needed, for example, of astronomical observations, as in the case of other laws of nature."[17] But Lobachevsky did not conduct physical experiments to test the ancient postulate about parallels. Instead, he analyzed radical alternative hypotheses that one might make.

At the time, the University of Kazan was becoming an increasingly inhospitable place for innovation in science and mathematics. In 1819, a reactionary bureaucrat, Mikhail Magnitsky, argued that the university should be "publicly destroyed"; he complained that "the professors of the godless universities transmit to the unfortunate youth the fine poison of unbelief and of hatred for the lawful authorities."[18] Yet Alexander the Tsar preferred to improve rather than obliterate the university. So in 1820 Magnitsky, of all people, was appointed to the Ministry of Education to oversee the university as curator of Kazan. Magnitsky implemented reforms and censorship to oppose subversive free thinking, to instead base education closely on biblical teachings. While accusing and constraining professors, Magnitsky also made extravagant proposals, such as that the Pythagorean theorem should not be taught except as a theological proof of the Holy Trinity: two sides representing the Father and Son and the hypotenuse denoting God's love manifested onto humans by the Holy Ghost.[19]

In Kazan, Lobachevsky spent substantial effort defending his colleagues against Magnitsky in the early 1820s. At the same time, Lobachevsky also worked on an abstract revolutionary activity that few could hardly imagine. He investigated what would happen to geometry if he did *not* assume the traditional postulate about parallel lines. He proposed the following construction, on a plane. Consider two straight lines, A and B, connected by a line CD perpendicular to both. Clearly, some straight lines, such as E, which cross line A through point C, also intersect line B. Lobachevsky called all such lines "cutting lines." He expected, as usual, that line A itself does not cut line B. But he also imagined that there are other lines that also do not cut B although they intersect line *A* at point *C.* Consider *F* as being one such "noncutting line." Look at it.

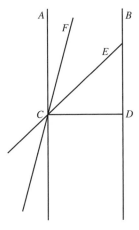

Figure 10.7. Lobachevsky's construction: if we extend lines *F* and *B*, will they meet?

Imagine that line *B* is extended upward. And imagine that line *F* is also extended. Does it seem like line *F* will cut line *B*? But Lobachevsky expected that it will *not* cut line *B*. By this point one might think that this is absurd. You can almost *see* that line *F* will cut line *B*.

But suppose that the diagram is not a clear representation of the situation. Suppose, for instance, that the slanting gap between lines *A* and *F* is not as big as pictured. On the plane, geometers would expect that only *one* line at *C* does not cross line *B*. But Lobachevsky was asking: How do we know that there is *only one* such line? What if instead there were other straight lines at *C* that do not ever cut line *B*, no matter how long they are extended? Lobachevsky then assumed that such noncutting lines other than *A* might exist.

But one might still be annoyed. If we accept what Lobachevsky assumed, then can't we just as well assume *anything*? Maybe not. The point is that Lobachevsky showed that despite his strange assumption, he could formulate a system of geometry that did not exhibit any contradictions. Still, we might think that if *F* is extended indefinitely, it *will* cut *B*. So perhaps it would help to somehow visualize some way in which no such cutting will happen. Consider the following example.

Suppose that you are painting the walls in your kitchen yellow, and once you're done you notice a small glob of paint on the kitchen floor. You look at it very closely, and you see that it has begun to dry, and its surface has a few wrinkles on it, as pictured.

Figure 10.8. A glob of paint.

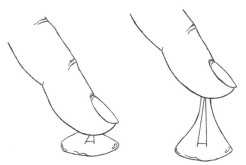

Figure 10.9. As the lines stretch, they seem to approach one another, will they meet?

One of the lines seems to lean toward the other, as it they would touch if extended. And for some reason, you put the tip of your index finger on it, and you find that the paint is still sticky: it sticks to your finger.

Now you pull back your finger and see that the glob of paint stretches upward, and you see that the lines stretch upward too. Okay, now imagine that you keep pulling your finger upward, *infinitely.* If we suppose that the paint keeps stretching, then you can imagine that the two lines on the paint stretch too and that the one seems to lean toward the other, yet they never meet.

Lobachevsky did not give any such example, but at least it illustrates a sense in which two lines might seem to approach one another but never meet. Another way to think about it is as follows.

Playfair and company did not *prove* that there is only *one* straight line through a point C that does not cross line B on the plane. That's why they called it a postulate, because nobody had proven that there must exist only one noncutting line. Hence, Lobachevsky proceeded to analyze what geometry becomes once we do not assume that traditional parallel postulate. He analyzed the consequences of there being maybe many such lines, even infinitely many.

One might expect that Lobachevsky then called all such lines "parallel lines," along with line A. But he didn't. Instead, he distinguished between the cutting lines and the noncutting lines, and he reasoned that there is a boundary line between those two kinds of lines. And so he called that boundary line a parallel line. Thus, he added, for any point C there would be not one but two "parallel" lines, P_1 and P_2, as illustrated.

Again, we might be annoyed. Why not call those two lines *boundary* lines, rather than calling them *parallel* to B? Having denied that there is only one line through C parallel to B, Lobachevsky was free to rescue the word *parallel* to use it to name whichever lines he wanted. He could have used it for line A, at the right angle from CD, but he didn't. Or he could have used it for the indefinitely many noncutting lines, but he didn't. Instead he chose to call the two boundary lines "parallel."

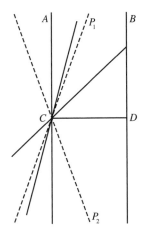

Figure 10.10. Lobachevsky's parallel lines, P_1 and P_2.

Next, Lobachevsky left the equal angles P_1CD and P_2CD unspecified, so that he admitted multiple possibilities. Given such arbitrary "angles of parallelism," there would be the two parallel lines, encompassing the infinitely many noncutting lines, such as F. Or, in particular, if such angles happen to be right angles (90 degrees), then the two parallel lines (and all the noncutting lines) merge into one, and we obtain Euclidean geometry. Therefore, Lobachevsky argued that Euclid's geometry was just a special limited subset of his more general geometry.

But we might yet disagree with him. After all, traditional geometry required that there be *only one* straight line parallel to B through a point. And Lobachevsky's geometry does not require this. There are other reasons why we might not agree that Euclidean geometry is compatible with Lobachevsky's geometry, but one example suffices.

Regardless, Lobachevsky showed that many theorems that are true in the ancient geometry are also true in his imaginary geometry. For example, the *Elements* proves a Proposition 16 that states that in any triangle, if one side is extended to make an exterior angle *e*, as illustrated, then that exterior angle is greater than the two opposite interior angles, *a* and *b*.

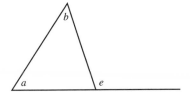

Figure 10.11. Elements, Proposition 16: for any triangle, $e > a$, and $e > b$.

This proposition, Lobachevsky showed, is also true in his imaginary geometry because it does not require Euclid's parallel postulate; it involves only postulates that are common to both geometries.

Also, Lobachevsky derived many geometrical propositions that were different, less narrow than the propositions proven in the *Elements* centuries before. For example, the *Elements* proved that the sum of the angles in any triangle makes two right angles or, as we now say, that all such sums make exactly 180 degrees. Instead, Lobachevsky demonstrated that in his more general geometry the sum of the angles in any given triangle makes two right angles *or less,* that is, 180 degrees *or less.* Weird, huh? Let's demonstrate this one proposition, as just one example of how Lobachevsky argued.

Consider a triangle *ABC,* and assume the counterintuitive claim that the sum of the angles is *greater* than 180 degrees. Lobachevsky proceeded to show that this is impossible, by arguing as follows. Take the smallest side, *BC,* and cut it in half at *D* with a straight line from *A,* and extend that line out to *E* by a length equal to *AD.*

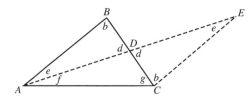

Figure 10.12. Given one triangle, we construct another triangle (dashed lines), the sum of their angles is equal.

Having assumed that the sum of the angles of the triangle *ABC* is 180 + *a,* Lobachevsky showed that the sum of the angles of the triangle *ACE* would also be 180 + *a.* Because the triangles *BDA* and *CDE* are congruent (you can see that their sides and angles are equal), such that the angle *ABD* is equal to the angle *ECD* (so they are both labeled *b*), and the angles *BAD* and *CED* are equal (so they are both labeled *e*), thus, since the sum of the angles of the triangle *ABC* is

$$b + e + f + g = (180 + a)$$

and the sum of the angles of the triangle *ACE* is also *b* + *e* + *f* + *g,* then the angles of *ACE* are indeed also equal to 180 + *a.* So what? The two triangles have the same angle sum, fine. But now Lobachevsky repeats the procedure upon the triangle *ACE,* constructing another triangle, as before, by cutting the middle of the shortest side *EC,* as pictured.

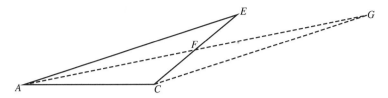

Figure 10.13. Again, we construct another triangle with a more obtuse angle, and the sum of the angles is the same as before.

Again, if we repeat our analysis, we can find that the sum of the angles of the new triangle *AGC* is also equal to 180 + *a*. But notice that the angle at G is very narrow, so Lobachevsky realized that he could continue to repeat the construction of triangle upon triangle, making ever-narrower angles, such that, eventually, one such triangle will have two angles that are each *less* than 1/2*a*. If so, then the sum of those two angles is *less* than *a*, so the remaining angle alone would then need to be *greater* than 180 degrees in order for the sum of the angles to be, as in all the previous triangles, 180 + *a*. But one angle alone just cannot be equal to 180 degrees (because then it is just a line; it cannot be the corner in a triangle), nor can it be greater than 180 degrees. Thus we reach a contradiction, an impossibility. Therefore, Lobachevsky concluded that the sum of the angles for any triangle *cannot* be greater than two right angles; it *cannot* be greater than 180 degrees.

So what? Brace yourself. Lobachevsky concluded not merely that the sum cannot be greater but instead that it can therefore be equal *or less* than 180 degrees. Since it cannot be true that 180 + *a* > 180, he concluded that *a* can be zero or *negative,* so that

180 + *a* ≤ 180

Against tradition, Lobachevsky said that triangles having angles that add to less than two right angles could exist. But he stopped short of asserting their existence. He did not expect that *some* triangles might exist having angle measures of less than 180 degrees alongside *other* triangles summing to exactly 180 degrees. Instead, he expected that if even *one* triangle has angles that add up to 180 degrees, then *all* triangles have that same property. Conversely, he also allowed that maybe no such traditional triangles exist.

Lobachevsky stated that while the usual angle sums of triangles led to Euclid's geometry, the alternative lesser angle sums led to "a new geometric science" that he called "Imaginary geometry."

But what do *you* think? Do you think that Euclid's triangles exist and that Lobachevsky's triangles do not? Does it seem that Lobachevsky's triangles are just nonsense? There might be a temptation to think that Euclid's are *real* triangles and that the rest are just imaginary. At universities, I am often stunned to meet students who believe that they have *seen* real triangles, that triangular drawings *really are* triangles. Their teachers do not teach them what ancient geometers knew: that lines of ink are not mathematical straight lines. For example, at around 300 CE, Porphyry explained: "As the geometricians cannot express incorporeal forms in words, and have recourse to the drawings of figures, saying 'This is a triangle,' and yet do not mean that the actually seen lines are *the* triangle, but only what they represent, the knowledge in the mind, so the Pythagoreans used the same objective method in respect to the reasons and forms."[20] Ink diagrams are just representations of triangles. A plastic triangular object is not a triangle either. None of us has ever *seen* a Euclidean triangle: three perfectly straight lines having length but no width, all on a perfectly flat plane, connected at three points, making angles that add up to exactly 180.00000000000000000000 . . . degrees. We cannot buy that at a store. Are Euclid's triangles myths? Like unicorns?

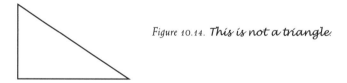

Figure 10.14. *This is not a triangle.*

Even if we consider three points in space, what if the straightest possible lines that can really exist between them do not make angles that add up to exactly two right angles? Do Euclid's triangles exist in physical space?

Legend says that Gauss tested the physical validity of Euclid's triangles by directing observers at three distant mountaintops to each measure whether 180 degrees was the sum of their angles.[21] But as shown by historians, this story is a myth.[22] From 1821 to 1825, Gauss carried out geodetic measurements of Hanover to help link the lands of Prussia and Denmark. His team made measurements from hills, mountain peaks, and church steeples. But in his geodetic labors Gauss *presupposed* the validity of Euclidean geometry, using it to check the accuracy of his observations. Later, reportedly, Gauss commented that geodetic measurements show that Euclid's parallel postulate is at least approximately valid.[23] So he calculated and realized that his measurements were fairly consonant with ancient geometry, although he had not carried them out as a test of any geometry. Another hearsay report, from much later, claimed that Gauss wondered whether maybe measurements of starlight might serve to test the apparent empirical validity of a triangle having 180 degrees.[24]

This brings us back to Lobachevsky's original motivation. He had become convinced that only physical experiments, astronomical observations, could possibly decide whether Euclid's postulates are true. Lobachevsky analyzed the parallax of starlight as the Earth orbits the Sun, but the data was inconclusive. Lacking any such decisive scientific proof, he proceeded to explore the broader, plausible geometry that included traditional structures as just one possibility.

From the 1820s until 1840, Lobachevsky published various accounts of his strange geometry. And in the new geometry János Bolyai solved an impossible famous problem of ancient geometry: he squared the circle (he found a rectangular area equal to the area of a circle).[25] Mathematicians, however, disregarded their works. In letters, the famous Gauss privately complimented Bolyai and Lobachevsky, but Gauss did not advertise them. Their incredible achievements remained unknown. Bolyai became embittered that Gauss kept the new geometry concealed, that he chose to fail at the moral obligation to awaken lethargic mathematicians to the existence and importance of the new geometry.[26]

In Russia, the troublesome Magnitsky was dismissed as curator of Kazan in 1826, for mishandling state monies, and subsequently Lobachevsky became rector of the University. He served successfully from 1827 until 1846. Afterward his health failed, and he became blind.

Meanwhile, Bolyai became an officer for the Austrian army. He often fell sick, with malaria, cholera, and rheumatic diseases. Yet Bolyai was temperamental and fought duels with his sword. In 1831, his father described János as "a virtuoso with the violin, good at fencing and brave," and noted that he "has often dueled, and overall is fierce as a soldier—but also a very fine light in the darkness—and darkness in light, and a passionate mathematician."[27] Reportedly, he was a passionate violin player and a highly skilled swordsman, and one time, "in a garrison with cavalry regiment, Bolyai was challenged by thirteen officers, which challenge he accepted on the condition that after each duel he be allowed to play a piece on his violin. Against all thirteen of his opponents, in these duels he emerged victorious."[28] One biographer said: "Bolyai was a compound of Saladin and Richard, fighting with a Damascus blade which cut silken cushions or chopped iron. Franz Schmidt told me in

Budapesth that his father had seen Bolyai lop off a spike driven into his door-post, and that some of his duels were to the death."[29] Thus Bolyai, who saw himself as "Euclid's phoenix," self-sacrificed and reborn from fire to "bless the human race," gained notoriety during his lifetime for drawing blood instead of gaining fame among mathematicians.[30]

The cold reception given to the innovative works of Bolyai and Lobachevsky reminds me of the very first lines of the translator's preface to the King James Bible, first published in 1611, nowadays often omitted, which in one later rendering states:

> **The best things have been slandered.**
> Any effort to promote the common good, whether by creating something ourselves, or by adapting the work of others, surely deserves serious respect and consideration, yet it finds only a cold reception in the world. It is greeted with suspicion instead of interest, and with disparagement instead of gratitude. And if there is any room for quibbling (and quibblers will invent a pretext if they do not find one), it is sure to be misrepresented and risk being condemned. Was anything ever undertaken that was new or improved in any way that did not run into storms of criticism?[31]

Accordingly, Gauss chose to keep his own work on the seemingly impossible geometry relatively secret.

Concerns for the physical validity of Euclid's geometry led another mathematician to develop more new geometries. Bernhard Riemann learned from Gauss, his professor, that Euclid's geometry might not be exactly applicable to the physical world. Gauss doubted the very truth of traditional geometry. Accordingly, Riemann suspected that Euclid and others had based geometry on generalizations from physical experiences, rather than on certainties.

Riemann expected that some properties of physical space can only be ascertained by observations and experiments. Other properties would be presupposed by our notions of space. In 1854, at Göttingen University, he delivered a lecture titled "On the Hypotheses That Lay at the Foundation of Geometry."[32] By deriving the logical properties of space on the basis of our fundamental notions, Riemann argued that Euclid's principles, after all, were not self-evident truths but were actually physical hypotheses.

For example, Riemann wondered, what if physical space is not infinite? We have no evidence that it *is* infinite, so we should well consider a geometry that does not involve that presupposition. Riemann developed such a geometry, based on lines that are not infinite and where parallel lines do not even exist. Riemann showed that given one straight line, and a point outside it, there might well be no line through that point parallel to the first. In that case, he showed, any straight line intersects any other straight line at *two* points. And all lines perpendicular to one line would meet at a point. He also showed that then the angles of any one triangle add up to *more* than 180 degrees. Moreover, in Riemann's geometry, Proposition 16 from the *Elements* (that the exterior angle in a triangle is greater than each opposite interior angle) is false.

None of this might seem to make sense. Didn't we just prove, at least, that the angles of a triangle *cannot* be greater than 180 degrees? Well, Lobachevsky's arguments depend on certain assumptions—whereas Riemann proposed distinct assumptions and analyzed their consequences. Moreover, Riemann knew that his propositions were not as incredible as they might sound at first, because he realized that the geometry with no parallel lines is comparable to the geometry of a sphere. Consider the great circles along the surface of a sphere, three of which are illustrated. Each of them intersects another at two points. Riemann expected this same property for his finite straight lines, as mentioned above.

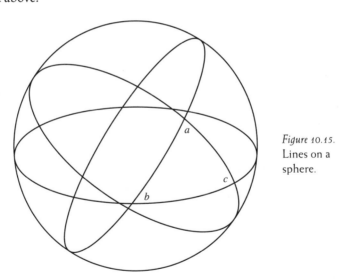

Figure 10.15. Lines on a sphere.

Likewise, the angles *a, b, c* on a "triangle" made by the great circles add up to more than 180 degrees. The same would take place with Riemann's finite straight lines.

None of this means that Riemann's geometry is the same thing as the geometry of a sphere—it isn't. Even the ancient geometers were well acquainted with the properties of spheres. Instead, the point is that since key properties of Riemann's strange geometry can be represented by analogies to the surface of a sphere, Riemann's geometry is not unthinkable. Just as there are no parallel lines on a sphere, there are none in Riemann's geometry. Likewise, the geometry of physical space might involve no such parallel lines at all, only things that *resemble* parallel lines. The wooden boards on the floor are not perfectly parallel; neither are the tiles on the bathroom wall. The point is that physical space might have a consistent geometrical structure even if parallel lines do not exist. Moreover, Riemann showed how to imagine infinitely many geometries.

Gauss died in 1855, having published no work on the "anti-Euclidean" geometry. Lobachevsky died in 1856, depressed, without knowing of Riemann's recent work. Farkas Bolyai too died in 1856, and his son in 1860, both demoralized and isolated. As for Riemann, he suffered from pleurisy and illness and died in 1866, aged thirty-nine.

Gauss's manuscripts were found after his death, stirring interest in non-Euclidean ideas. Mathematicians had ignored Bolyai and Lobachevsky, but Gauss was an acknowledged master, so they became curious. And Riemann's lecture, published in 1867, also drew attention.

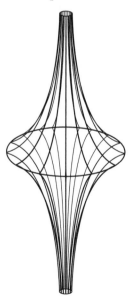

In 1868, Eugenio Beltrami published *Essay on an Interpretation of Non-Euclidean Geometry*.[33] Beltrami argued that just as one of Riemann's geometries could be related to the properties of great circles on a sphere, the geometry of Lobachevsky (and of Gauss and Bolyai) also could be interpreted in terms of Euclidean geometry. Beltrami had found that for a surface called a pseudo-sphere (figure 10.16) the propositions of Lobachevsky could be modeled.

The pseudo-sphere is supposed to extend infinitely in opposite directions, and thus certain lines approach one another but never meet. Many lines on that surface can cross a point and yet never cut a nearby line. Beltrami thus showed that Lobachevsky's geometry was logically consistent insofar as its theorems could be modeled in Euclidean geometry. Henri Poincaré concluded that

Figure 10.16.
Pseudo-sphere.

if a falsehood were found in any such theorem, it would falsify not only the strange new geometry but also traditional geometry.

The non-Euclidean geometries shocked many people. Not only were there new geometries, but it was not clear which one of them was physically true. Perhaps none of them was true. People had mistakenly believed that Euclid's principles were self-evident, universal truths. Instead it became apparent that such principles were based on everyday experience. Still, many mathematicians continued to think that the one true geometry was Euclid's and that the new geometries were logical fictions.

But by 1919 such reservations eroded. Astronomers found that a new theory of gravity successfully described the paths of light across great distances. Einstein had formulated this theory by using Riemann's geometry.

In 1907, Einstein had realized that gravity might be a relative effect. He thought: if a man falls from a rooftop, while he is falling he will not feel his own weight. For example, if you're falling while you hold a hammer in your hand, the hammer, as you fall, will not feel heavy in your hand. And if you release the hammer in midair, it will not seem to fall relative to you but will seem to just float next to you—that is, unless you look at your surroundings or hit the ground. Thus, a falling person would not feel the gravity that we are normally experiencing.

Moreover, if a person were inside a box, in outer space, beyond the Earth's gravity, while this box accelerates upward at 32 feet per second, then the person inside the box would actually feel the apparent effects of gravity. The person would feel heavy, stuck to the floor of the box. Any handheld object, once released, would seem to fall toward the floor, even if there were no Earth underneath. Thus, Einstein realized, the effects of acceleration would be identical to the effects of gravity.

How does this relate to non-Euclidean geometry? Well, people normally assumed that light travels in straight lines. But imagine now that light enters through a hole in the wall of Einstein's accelerating box. Then, since the floor rushes upward, the observer inside would see that the beam of light curves downward, as if it too were heavy, as if it were tending to fall down.

Einstein realized that if gravity and acceleration were truly equivalent, then this effect of curving light should also take place if light were traveling near a huge body. In particular, starlight should be attracted by the Sun: it should bend away from a straight path, as if falling toward the Sun.

Einstein hence formulated a theory of gravity in which immense bodies bend the space around them, noticeably, such that light would not travel in

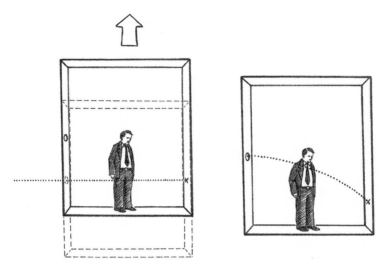

Figure 10.17. Left, according to people outside the box, the box accelerates upward and a light ray travels in a straight line through a hole in the box. *Right,* according to someone inside the box, the box is not accelerating at all, but there is gravity inside, and light entering through a hole falls toward the floor.

straight paths. Then in 1919, British astronomers in Brazil and West Africa observed a solar eclipse to test whether starlight passing near the Sun was really deflected, as Einstein predicted. The Moon blocked sunlight sufficiently so that several stars near the Sun could be observed with telescopes and photographed. Their positions were thus measured, and hence the astronomers

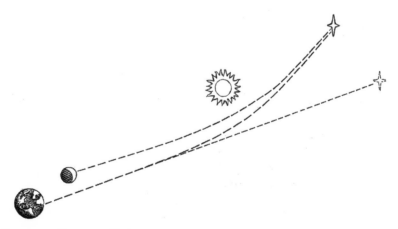

Figure 10.18. Two rays of light are emitted by a distant star. Both rays are drawn by the Sun, such that one ray does not reach us, but the other does. The ray that reaches us produces the optical impression that the star is displaced away from the Sun.

concluded that starlight had curved, and hence space had bent, pretty much as Einstein had predicted.

By the way, some philosophers have argued that in judging the eclipse photographs, the leading British astronomer was so biased in favor of Einstein's theory that he misrepresented the data.[34] This story has become a recent favorite among writers who debunk myths about the history of science.[35] However, historian and physicist Daniel Kennefick now has meticulously shown that the original data analysis of 1919 was actually fair.[36]

The reactions to the eclipse observations were extraordinary, partly because of the way in which the event was portrayed in newspapers. Reportedly, the president of the Royal Society of London, Sir J. J. Thomson, declared that Einstein's theory was "one of the greatest—perhaps the greatest—of achievements in the history of human thought."[37] New York Times reporters claimed that Einstein's theory was "A Book for 12 Wise Men," an esoteric work that could hardly be understood.[38] Reportedly, Thomson said that "it is not possible to put Einstein's theory into really intelligible words" and that "the difference between the theories of Newton and those of Einstein were infinitesimal in a popular sense," but that it was a discovery of the greatest importance to science.[39] According to the reporters, an astronomer from the eclipse expedition remarked that the discovery showed "that two lines normally known to be parallel do meet eventually, that a circle is not really circular, that three angles of a triangle do not necessarily make the sum total of two right angles."

The public reaction was staggering: people hailed Einstein for having replaced Newton's seemingly perfect physics and for having challenged Euclid's geometry. Einstein suffocated: "Since the result of the deflection of starlight became public, such a cult for me has arisen that I feel like a pagan idol."[40] He complained: "Like the man in the fairy tale who turned everything into gold, what he touched, so with me everything turns into newspaper hype."[41]

Nevertheless, Einstein became so impressed by the utility of non-Euclidean geometry that he concluded that ancient philosophers and mathematicians had been right: pure thought can comprehend the structure of nature. Accordingly, writers came to increasingly compare Einstein's achievements with those of the ancients: "Einstein, like Copernicus, was close to the ancient Pythagoreans. His years of struggling with general relativity taught him a soaring respect for the power of mathematics."[42] But notice, if one kind of geometry indeed had succeeded, another had failed. Physicists argued that the traditional space of Euclid, of which the "characteristic property of this space is that stated by the theorem of Pythagoras," had limited physical validity.[43]

Apparently space itself was curved. But years earlier, Poincaré had proposed another outlook: that if light rays are not straight, we may still suppose that space is Euclidean and that only the rays are curved. Thus the geometry employed in physics is chosen by convention. This viewpoint was somewhat echoed by someone at the meeting of the Royal Society when the eclipse observations were announced. The reporters noted that one of the speakers "suggested that Euclid was knocked out. Schoolboys should not rejoice prematurely, for it is pointed out that Euclid laid down the axiom that parallel straight lines, if produced ever so far, would not meet. He said nothing about light lines."[44] But most physicists did not accept Poincaré's thoughtful outlook. Einstein's theory of gravity convinced them that the geometry of physical space is dynamic: mass changes the structure of space.

If three rays of light traveling across immense distances determine a closed figure with three corners, its three angles would not measure 180 degrees. Does that mean that Euclid's triangles do not exist? And if so, is it fair to use the word *triangle* for the figures that roughly resemble Euclid's triangles? That is what physicists often do nowadays when they refer to a "triangle" in non-Euclidean geometries or in physical space. But we might still feel uncomfortable, because another alternative would be to give a new name to the three-angled non-Euclidean figure and to reserve the word *triangle* for the traditional concept, even if such a thing does not physically exist.

One might still imagine that in a perfectly empty space, far away from all objects that exert any gravity, light travels in perfectly straight Euclidean lines. But again, what if no such perfectly empty space exists? Physicists increasingly realized that even the vacuum, space with no air, is seething with subatomic particles and radiation, subject to gravitational effects.

Consider again the surface of a sphere. There are no straight lines there, none in the old sense of the word *straight,* and no plane triangles either. Yet there can be many other figures on that curved surface. Well, what if the world is like that? Even if there are no such things as "perfectly" straight lines or planes, physically there can be many other shapes, like the shapes that we encounter every day, where the edges of objects are only roughly straight. Thus, even if the elements of Euclid's geometry do not exist physically, there might exist a different geometry, a *physical geometry.* One can try to find its properties, instead of just assuming that the one true geometry is what we learned in school.

We might live without Euclid's triangles, but what about spheres? For centuries philosophers and astronomers and mathematicians sought circles and spheres in the heavens. Plato claimed that God "made the universe a circle moving in a circle, one and solitary."[45] Copernicus argued that the seemingly irregular motions of the planets follow a constant law wherein their motions are necessarily composed of circles.[46] He claimed that the heavenly world is spherical "because this figure is the most perfect of all."[47] Kepler envisioned the orbits of the planets as being set by spheres interspersed among the five Platonic figures. Meanwhile, one heretic did awake from the spell of heavenly geometry. Unlike the rest, the unrepentant martyr for free expression disagreed, Giordano Bruno denied the circles in the sky. Bruno said that "of the motions that we see sensibly and physically in natural bodies, there is none that in a great way does not differ from the simply circular."[48] He said that no body is truly spherical and that the motions of planets are not circular, but more like spirals.

And Kepler advanced beyond conjecture by mathematically analyzing the trajectories of the planets. In particular, Kepler struggled for years analyzing Tycho Brahe's observations on the orbit of Mars. It was Kepler's war. By 1607, he published his findings: the Pythagoreans, Ptolemy, Brahe, Copernicus, and nearly everyone else were wrong. The orbit of Mars was not circular; it was nearly elliptical. The Sun was not at that orbit's center but at one of the focal points of the ellipse. The same held for other planets. It was a huge discovery, but when Galileo heard that Kepler showed that the orbits were not circular, he just disregarded it.[49] Like other astronomers, Galileo seemed engulfed by the idea that the structure of the heavens is perfectly geometrical. Likewise, Galileo did not believe Brahe's discovery that comets travel above the Moon's orbit. Why not? Partly because the paths of comets were not circular, and it seemed impossible to Galileo that heavenly bodies could move in a noncircular motion. He therefore concluded, wrongly, that comets are not objects at all, that they are just optical illusions in the sky. And he attributed that notion to Pythagoras.[50]

Nowadays many people think that in free space objects obey the law of inertia: that objects move uniformly in a straight line. It's an old myth that such a law of inertia was proposed by Galileo. For him, all motions in the heavens had to be circular; he *denied* the contrary. For example, in contrast to Paolo Antonio Foscarini's rejection of eccentrics and epicycles, Galileo insisted that "they undoubtedly exist in the heavens," in accord with Copernicus's scheme.[51] In 1624 Galileo wrote: "I tell you that if natural bodies have it from Nature to be moved by any movement, this can only be circular motion, nor is it possible that Nature has given to any of its integral bodies a propensity to be moved by straight motion. I have many confirmations of this proposition."[52] Galileo even argued that the primary motions of animals are all circular.[53]

Kepler too relished this illusion, since he continued to look for Pythagorean harmonies and Platonic solids. After having realized that the orbits were nearly elliptical, he still expected the planets to be separated by spheres and the five regular solids.[54] To him, geometry was coextensive with religion. Kepler argued that geometry "is coeternal with God" and that it shone forth in the divine mind, providing patterns to furnish the world.[55] He even tried to prove that the entire universe is a sphere.

Scientists eventually dismissed Kepler's scheme of spheres and solids: "that whole Mystery is nothing but an idle Dream taken from Pythagoras or Plato's Philosophy."[56] None of what astronomers saw was really circular or spherical; the Moon is not a sphere, orbits are not circular, and so on. Did it end there? Were we finally freed of the compulsion to imagine circles in the fundamental structure of the universe?

One day, while teaching a class, I was looking at a science textbook, and suddenly I was stunned to realize that the circles we criticized in ancient astronomy had migrated into microphysics, almost without objections. Chemistry books show molecules as made of spheres. Atoms have been portrayed as spheres for centuries, but we've known for a long time that they have structure, parts. So now chemistry and physics books show atoms, composed of what? Spheres again. Protons and neutrons appear as little spheres. Electrons seem to orbit the nucleus in circular paths, and in nearly all books they're portrayed as spheres.[57] Yet physicists have found that protons and neutrons have parts, so how do they portray them? Spheres again!

Repeatedly, the entire universe seems to be made of invisible spheres. Moreover, according to wave theory, light propagates in expanding spheres. And the spherical wave fronts are themselves composed of smaller expanding spheres. And a plane wave front, a flat surface, even that is imagined as composed of the sum of many expanding spheres.

So we no longer think of planets as spheres moving in circles, but people still imagine things as composed of spheres. Such imagery is accepted as just useful models that approximately describe things. But maybe it's worse than that. Apparently *every time* scientists have thought that a physical structure was spherical, it actually was not. The image of a sphere was a simplifying fiction, a pleasant stand-in for the real structure. But often people took that image as real. Why? Was it unpleasant to imagine asymmetries? For centuries, neatly round images stood in the way, like roadblocks, repeatedly distracting scientists from trying to find the real structure of things.

It would be spectacular to actually find a perfect sphere, anywhere. It would be like finding a mythical creature, a unicorn. Circles often function like myths, as figures that disguise our ignorance. Should we stop representing subatomic things in terms of circles? Maybe use something uglier, any interesting, annoying shape that triggers curiosity? For example, when Werner Heisenberg was a young student of physics, he suffered a textbook that illustrated atoms as including little hooks connecting one to another.[58] He felt annoyed by that idea, he thought that atoms could not possibly have that shape, and so eventually he investigated atomic physics. But how many students nowadays feel annoyed when they see a diagram of atoms and other particles, all spherical?

When individuals have replaced circles with more realistic shapes, they have been celebrated for advancing science. So, whenever a physics book depicts some structure as a circle, we may interpret it as a sign that says, "This is not the structure." Thus Newton once drew the Philosophers' Stone as a set of circles. And consider the ancient claim, "Pythagoras believed that time is

Figure 10.19. Are circles less imaginary than unicorns?

the encompassing sphere."[59] It fits the habit of pretending to understand by using beguiling round imagery.

Many people still say that "mathematics is the language of nature." But we should ask: "*Which one?*" Because there isn't just one mathematics; there are many—and they're not equivalent; they don't all say the same thing. Galileo said that the book of nature is written in the language of mathematics: triangles and circles.[60] Yet we have no evidence that such things exist, physically. How about: "English is the language of nature"? Why not say this? Because there are other languages, of course, so no matter how well English describes nature, it's just one language. Some kinds of mathematics describe physical processes well, but altogether mathematics is not a language. It does not need to stand for anything. It stands for itself.

As for usefulness, for many mathematicians it's of no concern. The English mathematician G. H. Hardy famously explained: "I have never done anything 'useful.' No discovery of mine has made, or is likely to make, directly or indirectly, for good or ill, the least difference to the amenity of the world. I have helped to train other mathematicians, but mathematicians of the same kind as myself, and their work has been, so far at any rate as I have helped them to it, as useless as my own. Judged by all practical standards, the value of my mathematical life is nil; and outside mathematics it is trivial anyhow. I have just one chance of escaping a verdict of complete triviality, that I may be judged to have created something worth creating. And that I have created something is undeniable: the question is about its value."[61]

History shows that we should not teach the elements of geometry as if they're obviously true. Just as Lobachevsky's triangles are bizarre, so too are Euclid's triangles. We have never seen perfectly flat figures made of three perfectly straight lines having no thickness and that meet in corners that add up to two right angles. Such things exist in our minds, but do they also exist outside? Such things can rule in our minds, by habit, but we're presumptuous to think that they rule over physical things.

11

INVENTING MATHEMATICS?

Through the so-called Platonist outlook, many people construed mathematics in religious ways. They assumed that its principles were eternal truths discovered by special men, geniuses, and they accepted that these truths were valid everywhere and could never change. The laws of geometry and numbers seemed like the laws of God, and therefore mathematics was valued as a preparation to discipline the mind for studies of metaphysics and theology. In the 1730s, Bishop Berkeley complained that some people accepted strange mathematical propositions on the basis of faith instead of reason. Some students learned rules on the basis of authority, "because the Master said it," like the Pythagoreans.

Along with rules of operation, students learned the rules of "Thou shalt not." Children learn that they *cannot* subtract a greater number from a lesser. Later they learn that actually it's possible. But they then learn that they cannot take square roots of negative numbers. Later, they learn that it is possible to take roots of negatives, and they learn about imaginary numbers, and so forth. As they advance in mathematics, they learn that a series of operations that first seemed impossible are actually fine.

Even nowadays, some students still learn certain basic rules by obedience. In an interview, when asked why we must use the rule that division by zero is undefined, one ninth-grade student replied: "The question is not 'Why is this the rule?' You just have to know the rule. Clever mathematicians make rules and we should memorize them. The problems we want to solve by applying the rules are what we have to understand." Another student, in the eleventh grade, said: "It is not allowed to divide by zero. In mathematics we have rules, and we operate according to them. These rules often do not seem reasonable. For instance, it is illogical that minus times minus is plus. When studying mathematics, we have to obey the rules and to work with them. There is no point at all in looking for explanations. One just has to accept them."[1]

To illustrate how people variously explain rules, consider negative numbers. When multiplying positives and negatives we have

$$+ \times + = +$$

$$- \times + = -$$

$$+ \times - = -$$

$$- \times - = +$$

Years ago, in 1992, a college freshman told me "why minus times minus is plus," by making + mean "good." She said:

> When good things happen to good people, that's good.
>
> When bad things happen to good people, that's bad.
>
> When good things happen to bad people, that's bad.
>
> When bad things happen to bad people, *that's good!*

I laughed. But we might think that if bad things happen to bad people, that's not really good. Maybe we imagine that if good things happen to bad people, they might cease to be bad.

The same problem affects some popular justifications for why the product of two negatives is positive. One claim is that, in language, two negatives make a positive. A sentence such as "I do not disagree" seems to express agreement. But it sounds ambiguous: maybe the speaker is just undecided. Likewise, when someone says, "Don't bring me no food," it might mean that he really does not want food. Still, some linguists tried to find certainty or universality in the way that positives and negatives are used in human languages. A famous anecdote says that in the 1950s, the British philosopher of language J. L. Austin, a professor at Oxford University, presented a lecture in which he argued that although double negatives often express positives, in many languages, there exists *no language in which two positives express a negative.* But then someone in the audience replied in a dismissive voice: "Professor Sidney Morgenbesser is said to have piped up from the back of the room with an instant, sarcastic, 'Yeah, yeah.' This convulsed the audience in laughter and put a blot on the speaker's career."[2]

The larger problem in claims that language entails that minus times minus is plus is that double negatives in language hardly involve *multiplication*. Two negatives might make a positive, but why not apply that, say, to addition rather than multiplication? If someone says, "I do not disagree," again, what exactly is being multiplied? Conventions of grammar do not decide mathematical rules.

Instead, many teachers try to justify minus times minus by using practical or physical analogies. Negatives are often used to represent debts. In 1770, Euler explained why a negative times a positive makes a negative by referring to debts. Say $-5 \times 3 = -15$ because a debt of five dollars times three makes a debt of fifteen dollars. However, Euler did *not* refer to debts to explain why minus times minus is plus. Likewise, teachers today do not say that $-4 \times -4 = 16$ because of money; it just doesn't make sense. A debt of four dollars multiplied by a debt of four dollars makes sixteen dollars? Or say, a debt of four dollars not taken four times makes a gain of sixteen dollars? Shouldn't it keep one's account *unchanged*? Why would it make a gain?

Negatives are also used to represent temperatures. But again, it makes no sense to multiply two negative temperatures together, making a positive temperature. So teachers do not use this analogy either. The point is that physical analogies or examples from daily life do not justify rules, because they only work in limited contexts, and only for some rules, not for all.

Algebraists reached this important conclusion long ago. For example, in 1805, the foremost professor of algebra in France, Sylvestre Lacroix, criticized how writers often used inconsistent analogies when trying to justify operations with isolated negatives: "Those who do not want to make it into a matter of authority have tried to explain the nature of such quantities, having made recourse to forced comparisons, like that of assets and debts, which are not convenient but only in particular cases." Lacroix also argued that geometry does not fully explain operations with negatives, because the theory of negatives really consists just of algebraic facts with which one should be content. He said: "In mathematics one abuses reason when one obstinately persists in not recognizing certain facts resulting from the combinations of calculations, which cannot be explained more clearly than by themselves."[3] Accordingly, to avoid inconsistencies, mathematicians choose to justify the traditional rules of signs not on any systematic physical analogy, but on abstract principles.

Thus, mathematicians justify "minus times minus" on the basis of formal rules. In particular, they derive or prove that minus times minus is plus by using the distributive rule of multiplication: $(a + b)c = ac + bc$. For example:

$$0 \times -1 = 0$$
$$(-1 + 1) \times -1 = 0$$
$$(-1 \times -1) + (1 \times -1) = 0$$
$$(-1 \times -1) + -1 = 0$$
$$-1 \times -1 = 1$$

So, mathematicians explain that "minus times minus is plus" is not an assumption; it is instead a theorem, a consequence of the fundamental rules of numbers. But like other proofs, the simple proof above depends on certain assumptions. In particular, it depends on the assumption that positive times negative makes negative:

$$+ \times + = +$$
$$- \times + = -$$
$$+ \times - = -$$

If we prefer to call these rules something other than "assumptions," then we need to posit other assumptions from which we can derive these rules. But even then, we are choosing assumptions in order to obtain preestablished results; we are choosing *rules that will lead to* minus times minus is plus.

Mathematicians know that algebra is based on rules of signs that are established partly by conventions. Yet we rarely experiment with modifying such conventions to see what different kinds of mathematics would result. Therefore, years ago, I experimented with the rules of signs, to see whether we can invent a scheme in which minus times minus is minus.[4] Instead of the three assumptions above, suppose that we change one of them:

$$+ \times + = +$$
$$- \times + = -$$
$$+ \times - = +$$

Here, multiplication is not commutative: $(- \times +) \neq (+ \times -)$. Centuries ago, this would have seemed impossible, but ever since William Rowan Hamilton formulated the algebra of quaternions, it became clear that there can exist mathematical concepts, even numbers, that do not obey the commutative rule of multiplication. To analyze negative numbers, most mathematicians had assumed that they had to operate with the same rules as positives. But now we know that not all numbers need to obey the same rules—what if mathematicians had known this when they first analyzed negative numbers? Could they have formulated rules of operation different from the rules of positives?

Suppose that minus times plus is minus, as usual, but suppose also that plus times minus is plus. We can then carry out a proof similar to the previous one:

$$0 \times -1 = 0$$
$$(-1 + 1) \times -1 = 0$$
$$(-1 \times -1) + (1 \times -1) = 0$$
$$(-1 \times -1) + 1 = 0$$
$$-1 \times -1 = -1$$

But here, minus times minus is minus. This rule has been derived from the three assumptions above.

An immediate objection might be: *You can't do that.* But having seen the various radical developments in the history of mathematics, it's clear that change *is* possible in the elements of mathematics. Still, a common reflex is to view such changes as some sort of a challenge, as if the traditional rules are being rejected. From that view, it seems senseless to try to change a rule as fundamental and unproblematic as minus times minus is plus. But no, history shows repeatedly that the creation of a new kind of mathematics does not entail that prior systems are false or should be abandoned. People continue to use traditional ancient geometry despite the invention of new geometries by Bolyai, Lobachevsky, and others.

But traditionally, there is an asymmetry in how people view distinct parts of mathematics. It is well known that new geometries and algebras can be formulated by specifying modified axioms or rules. Yet the study of numbers remains a fundamental field where the Platonist philosophy still prevails. The Austrian philosopher Ludwig Wittgenstein complained that our thoughts are saturated with the habit of treating "arithmetic as the natural history (mineralogy) of numbers."[5] People tend to think of numbers as natural objects, rather than as concepts. They think of numbers as entities that obey universal, timeless rules, which cannot be changed at will.

When new numbers are proposed and accepted, they are usually viewed as discoveries. They seem to be expansions of the number system. In addition to whole numbers and fractions, past mathematicians conceived irrational numbers; for example, they conceived the square root of 2 as a number that lies somewhere between 7/5 and 3/2. Likewise, the number system was expanded to include zero, an element lacking many properties of all other numbers. Next, the system was expanded to include negative numbers and, later, imaginary and complex numbers. And Hamilton conceived quaternions as a further expansion of the complex number system. Likewise, if we consider the real number line, it is interesting to envision where there might be new numbers. Outside the line, of course, but in the line itself we have, for example, two kinds of numbers adjacent to one another: zero and positives. Is there a gap between zero and the positives? Can anything fit between zero and the smallest positive number? Infinitesimals are numbers imagined for that purpose.

All of these developments were *expansions* of the number system. And sometimes, apparent algebraic laws changed, such as when $a + a > a$, *for every a*, stopped being true once zero was accepted as a number, or when $ab = ba$ stopped being valid for all numbers. But even then, it still could seem that the "new" numbers had not been invented, but had existed forever, waiting to be discovered, say, beneath the positives, or between the positives and zero, or outside the number line, or beyond infinity. The Platonist philosophy could accommodate the growth of numbers as discoveries of abstract objects that had quietly existed, unnoticed, in an eternal realm.

But disagreements arose whenever mathematicians proposed different properties for any number. Some said that zero could be a divisor; others said that it could not. Some said that zero had no sign; others said it was positive. Others now talk about a negative zero, distinct from positive zero. If zero is a unique number, then it cannot have contradictory properties. When someone thinks of zero as a symbol that can have only *one meaning,* and believes that it exists in the only possible mathematics, then disputes arise. Similarly, people fought over the concept of infinity. Some writers criticized others as confused or insane.

To treat numbers as unique, eternal, and unchangeable is a kind of number mysticism reminiscent of the legendary Pythagoreans. But there is no reason for it. There can well be a numerical system in which division by zero is undefined, and at the same time there can be another system in which it gives infinity. And there can be other systems in which division by zero gives other values. In one system, zero can have certain properties, and in another it can well have other properties. Some systems can be useful, practical, and others might be interesting as abstract schemes of symbols. In one algebra minus times minus is plus; in another it might be something else.

Many rules of algebra did not originate from present axioms. Instead, the ordinary rules came first, such as how to multiply negatives. Afterward, for centuries, mathematicians gradually formulated postulates that they posited as the foundation of numerical algebra. Some mathematicians believed that they were seeking the preexisting and eternal foundations of numerical algebra. But it is unwarranted to assume that such structures really existed eternally before they were articulated.

Only centuries after mathematicians had often used basic rules, such as $- \times - = +$, did they manage to articulate systems of postulates from which to derive elementary algebra. For example, by the early 1900s, some writers had formulated numerical algebra in terms of five postulates, as follows:

I. If a and b are any two elements in a field F, $a + b$ and ab are uniquely determined elements of F, and $b + a = a + b$, and $ba = ab$.

II. If a, b, c are any three elements of F, $(a + b) + c = a + (b + c)$, $(ab)c = a(bc)$, $a(b + c) = ab + ac$.

III. There exist in F two distinct elements, denoted by 0, 1 such that if a is any element of F, $a + 0 = a$, $a1 = a$ (whence $0 + a = a$, $1a = a$, by I).

IV. Whatever be the element a of F, there exists in F an element x such that $a + x = 0$ (whence $x + a = 0$ by I).

V. Whatever be the element a (distinct from 0) of F, there exists in F an element y such that $ay = 1$ (whence $ya = 1$, by I).[6]

Geometry had been based on five postulates, so some mathematicians similarly formulated algebra in terms of five postulates. The "elements" are numbers, and negative numbers are introduced in postulate IV. These postulates serve to derive elementary numerical algebra, including theorems such as $- \times - = +$. However, if instead we choose to derive a different rule, such as $- \times - = -$, we can formulate a different set of postulates. For example:

I. If a and b are any two elements of R, $a + b$ and ab are uniquely determined elements of R, and $b + a = a + b$.

II. If a, b, c are any three elements of R, $(a + b) + c = a + (b + c)$, $(ab)c = a(bc)$, $(a + b)c = ac + bc$.

III. There exist in R two distinct elements, denoted by 0, 1 such that if a is any element of R, $a + 0 = a$, $a1 = a$ (whence $0 + a = a$, by I).

IV. Whatever be the element a of R, there exists in R an element x such that $a + x = 0$ (whence $x + a = 0$ by I).

V. Whatever be the element a (distinct from 0) of R, there exists in R an element y such that $ya = 1$.

Here, multiplication is not commutative, but it is associative: $(ab)c = a(bc)$. The left distributive rule $c(a + b) = ca + cb$ is not stated because it is not generally valid in this scheme (it is valid for positives, but not for combinations of positives and negatives). However, the right distributive property is valid for all elements: $(a + b)c = ac + bc$. From IV, we define negative numbers as $x = -a$. From V, we define fractions as $y = {}^1/_a$, and we define division as $z \div a = z \times {}^1/_a$, where the quotient always has the sign of the dividend z, following the sign rules for multiplication. So we now have

$$+ \div + = +$$
$$- \div + = -$$
$$+ \div - = +$$
$$- \div - = -$$

Again, a likely reflex is to think that this all must be impossible, that the rules of real numbers cannot be changed, following Platonism. We might expect that some contradiction must arise; for example, given the artificial rules, we have: $2 \times -3 = 6$. This result seems strange, but does it lead to a contradiction? If we transpose 2 to the right side of the equation, it might seem that we would have: $-3 = 6/2$, and therefore: $-3 = 3$. But this problem is only apparent; it arises because the artificial postulates and new sign rules were not fully applied. Instead, by applying these invented rules we get

$$2 \times -3 = 6$$
$$(\tfrac{1}{2}) \times (2 \times -3) = (\tfrac{1}{2}) \times 6$$
$$(\tfrac{1}{2} \times 2) \times -3 = 3$$
$$1 \times -3 = 3$$
$$3 = 3$$

There is no contradiction: the five new axioms lead to consistent results.

Consider another example. In this system, imaginaries need not arise. The new postulates entail

$$\sqrt{-9} = -3$$

instead of an imaginary number. This result may seem strange and artificial, but certainly not nearly as strange as the now accepted result originally seemed to mathematicians:

$$\sqrt{-9} = \pm 3i$$

This kind of result originally met with vigorous, long-lasting critiques and ridicule: for being imaginary, for having a double solution, for not having an immediate geometrical or physical meaning, and so on.

By changing the axioms, some of the theorems in the new system are different from those of traditional numerical algebra; for example, we now have $- \times - = -$ and other deviant rules. Some of the signs of results are different. Also, it is important to note that the changes in axioms lead not just to changes in some signs; they also produce changes in numerical results. For example, find the value of a. By following the traditional rules we obtain

$$a = (8 \times -3) + (-4^2 + \sqrt{-25})$$
$$a = -24 + (16 \pm 5i)$$
$$a = -8 \pm 5i$$

By contrast, the new and artificial system produces instead

$$a = (8 \times -3) + (-4^2 + \sqrt{-25})$$
$$a = 24 + (-16 + -5)$$
$$a = 3$$

The results are entirely different. Does that mean that the new and artificial system is wrong? No, differences are evident too when we compare the geometries of Euclid and Lobachevsky or when we compare any distinct mathematical systems. By modifying traditional axioms, we make a system that generates some different results. But habit rebels against this; a result such as $-8 \pm 5i$, we think, *must* be more meaningful than $a = 3$. It must have more physical meaning. And indeed, in certain physical contexts, such as electrical engineering, a result such as $-8 + 5i$ can be clearly meaningful. But that is because for centuries mathematicians and physicists struggled to find contexts and applications in which certain combinations of algebraic symbols would be meaningful. A result such as 3 might seem meaningless, but mainly because no comparable effort has been carried out to associate the alternative procedure with a practical context. But regardless, as we've seen previously, mathematicians established that it is quite irrelevant whether a mathematical structure corresponds to a physical problem. Some parts of mathematics are meaningful in some contexts, and others are not, which is not a problem for mathematics itself.

Consider another example. Following the invented sign rules, we have $\sqrt{4}\sqrt{-9} = (2)(-3) = 6$ and also $\sqrt{4 \times -9} = \sqrt{36} = 6$. Similar examples show that in this invented system

$$\sqrt{a}\,\sqrt{b} = \sqrt{ab}$$

for any values of a and b, positive, negative, or zero. By contrast, traditional algebra, as usually understood, requires that this equation is not valid for negative numbers. Some algebraic equations that are valid in the invented algebra are not valid in traditional algebra. For example, table 11 gives some examples of such similarities and differences.

In the artificial system we obtain some elegant results that are absent in traditional algebra. But they come at a price: the traditional system includes convenient algebraic equations that are not valid for combinations of positive and negative numbers in the artificial system. Still, some of those equations can be replaced with others that achieve the same purpose. For example, the so-called Pythagorean theorem is not valid in the artificial system, say, if a is positive while b is negative. However, we can formulate an equivalent theorem using the artificial postulates, say: $1a^2 + 1b^2 = 1c^2$. Here by multiplying 1 we make explicit the operation of changing each value to positive, which also happens in the traditional system because its operation of squaring converts negatives into positives.

Again, the artificial rules lead to consistent results.[7] And in this system, there are no imaginary numbers, complex numbers, or quaternions. Now of course, I am not proposing that such numbers should be eliminated or that the traditional rules of signs should be replaced. I am only showing that it is

Table 11. A Selection of Algebraic Equations, to Compare and Contrast Traditional Mathematics and an Artificial System of Postulates with Different Rules of Signs

Traditional Algebra	Artificial Algebra
$\sqrt{a} = \pm r$	$\sqrt{a} = r$
$\sqrt[x]{a}$ = multiple values for each a	$\sqrt[x]{a}$ = one value for each a
$(\sqrt[x]{a})^x = a$, in some definitions	$(\sqrt[x]{a})^x = a$
$\sqrt[x]{a^x} = a$, in some definitions	$\sqrt[x]{a^x} = a$
$(-1)^x = -1$, if x is even, not odd	$(-1)^x = -1$
$-(a^x) \neq (-a)^x$	$-(a^x) = (-a)^x$
$-\sqrt{a} \neq \sqrt{-a}$	$-\sqrt{a} = \sqrt{-a}$
$\sqrt{a}\,\sqrt{b} = \sqrt{ab}$, except if a and b are negative numbers	$\sqrt{a}\,\sqrt{b} = \sqrt{ab}$
$a^2+b^2 = c^2$	$a^2+b^2 = c^2$, if a, b, c have the same sign
$(a+b)^2 = a^2+2ab+b^2$	$(a+b)^2 = a^2+ab2+b^2$ if a, b have the same sign

possible to invent new numerical systems in which many operations are just the same as usual, for example, 2 + 2 = 4, while other operations are quite distinct, and interesting.

So what does it mean to write −1 × −1 = −1? Does it mean that a traditional rule is false? No. Does it mean that the operation of multiplication has been redefined? Does it mean that here the "−1" symbols are misrepresentations of concepts that do not really stand for negative one? Meanings are established by definition, by group conventions, so mathematicians can reject an expression such as −1 × −1 = −1 at the very least on the grounds that the symbols used in it have predefined meanings, accepted virtually universally, and that in due regard to that, this expression is plainly incorrect. This might sound like a strong argument, but it crumbles in view of history. It is a historical fact that symbols such as "−1" or "×" have had variously evolving meanings over time. Therefore, we cannot appeal to *the* meaning of a symbol to decide an issue, because such meanings can be redefined. The same symbols can and have been used in various distinct and incompatible ways in various contexts.

For example, the same expression "straight line" has different meanings in ancient geometry and in the geometry of Lobachevsky. But again, while this is well known in geometry, as in nearly all realms of words, symbols, and definitions, in arithmetic there still reigns the mystical Pythagorean impression that numbers are immaterial objects that have unique and unchangeable properties. I'm arguing, instead, that the properties of some numbers are established by definitions.

Since creativity in the axioms of mathematics is seldom taught, any such tinkering might seem unnecessary, perverse. By imagining traditional mathematics as "natural," someone might ask: "Why should we wander blindly, when instead we can study and climb a spectacular mountain, see its beautiful sights, and discover new parts of it that no one has ever seen?" Actually, an architectural analogy seems better: "Why should we bother to build a small structure of bricks, when instead we can explore a beautiful and immense pyramid and study its ancient mysteries?" Why? Because we want to know how pyramids are constructed—because we're curious to see if we can build something else. We should explore this kind of playful invention; mathematics is too often taught as if there is no room for creativity in its elements, despite all history to the contrary.

In 1883, Georg Cantor extended the system of natural numbers to include infinite ordinal numbers. To do so, he argued that the essence of mathematics is its freedom:

I believe that in these principles there is no need to fear any danger for science, as it seems to many people; first the solely specified conditions within which the freedom to construct numbers can be exercised, such that it leaves very little leeway for arbitrariness, but then it also carries with each mathematical concept the necessary corrective in itself: if it is infertile or inexpedient, it shows it by its uselessness and soon then it will be abandoned, for lack of success. By contrast, it seems to me that any unnecessary restriction to mathematical research brings with it a much greater danger, and one so great, as really no justification can be drawn from the essence of science for that, because the *essence of mathematics* lies precisely in its *freedom.*[8]

Cantor formulated his set theory on the basis of one-to-one correspondence between the elements of sets, which led him to the notion that there are numbers that are larger than all finite numbers, what he called transfinite numbers.

Cantor imagined that the ordinal numbers 0, 1, 2, 3, . . . , are followed by "transfinite numbers": ω, $\omega + 1$, $\omega + 2$, . . . , appropriately using omega, the last letter of the Greek alphabet. These numbers are "infinite" inasmuch as they are beyond all finite numbers, larger than them. Transfinite numbers have very strange properties; for example, they violate the commutative rule of addition:

$$\omega + 1 \neq 1 + \omega$$

Cantor was a spirited mathematician, but he was also moody and melancholic. In 1884, he suffered a nervous breakdown and was hospitalized. He blamed this on friction with a former professor, Leopold Kronecker. Still, he soon continued to work. To describe the size (cardinality) of infinite sets, Cantor imagined other transfinite numbers that he later designated using the Hebrew letter aleph, $\aleph$. For example, by 1895 he used the symbol $\aleph_0$ to refer to the cardinal number of the set of all natural numbers. And again, these new numbers had strange properties such as

$$\aleph_0 + \aleph_0 = \aleph_0$$

Cantor worked on his radical mathematics for years, while he suffered from bouts of manic-depression. He was repeatedly hospitalized but continued to work in isolation. What mattered to him was not whether transfinite numbers had physical applications or obeyed all laws of other numbers; instead he cared about self-consistency. Still, his system seemed so bizarre to some prominent mathematicians that they rudely ridiculed it.

In Berlin, Leopold Kronecker entirely rejected his theory. A legend later developed, reminiscent of Socrates: that Kronecker accused Cantor of being "a corrupter of youth." But this is not true: there's no evidence that Kronecker ever said that. The claim was made by Arthur Schoenflies, a follower of Cantor, who assumed that Kronecker had that opinion.[9] Nonetheless, Cantor complained that Kronecker rejected his theory as made-up "humbug," having no real significance: "As long as I work scientifically, my works are systematically attacked by Kronecker and accused as empty fantasies, without real basis."[10]

Cantor did not think that he had invented transfinite numbers. He thought that such numbers were a discovery, a divine revelation: "I entertain no doubts as to the truth of the transfinites, which I have recognized with God's help and which, in their diversity, I have studied for more than twenty years."[11] While Cantor ridiculed infinitesimals as self-contradictory nonsense—*"paper numbers that belong in the trashcan!"*—he yet believed that his transfinite numbers were sacred revelations.[12]

From mathematics, Cantor turned to theology, finally envisioning himself as a servant and messenger of God. In 1896, he wrote: "From me, Christian philosophy will be offered for the first time the true theory of the infinite."[13] In the words of historian Joseph Dauben, "this was a strong form of Platonism"—it did not matter to Cantor whether his transfinite numbers were practically useful or could be found in some form in the physical world, because to him they existed at least as necessary possibilities in the eternal mind of God.[14]

The prominent French mathematician Charles Hermite also rejected Cantor's theory. His student Henri Poincaré later recalled "the horror that for some time it has inspired in certain individuals, such as Hermite," who died in 1901.[15] Poincaré described Hermite's complaints as follows:

> He readily repeated: I am anti-Cantor *because* I am a realist. He
> reproached Cantor for having created new objects, instead of
> contenting himself with discovering them. Doubtless because of his
> religious convictions he regarded it as a sort of impiety to want to
> penetrate flat-footed into a domain that God alone can grasp and for

not waiting for him to reveal his mysteries to us one by one. He [Hermite] compared the mathematical sciences to the natural sciences. A naturalist who sought to divine the secret of God, instead of consulting experience, would have seemed to him not only presumptuous but disrespectful of divine majesty.[16]

Cantor used the idea of "actual infinity." But Poincaré complained: "There exists no actual infinity, and when we speak of an infinite set, we mean to say a set onto which one can ceaselessly add new elements (similar to a subscription list that will never be closed to the addition of new subscribers)."[17] Poincaré argued that Cantor's set theory involved meaningless concepts and entailed paradoxes and contradictions, such that most of its concepts should be banished from respectable mathematics. Poincaré commented: "As for me, I think, and I'm not the only one, that the important point is to never introduce anything other than entities that one can define completely in a finite number of words. Whatever the remedy adopted may be, we can promise ourselves the joy of the medical doctor who is called to pursue a beautiful pathological case."[18] The doctor was fighting an illness: infinity.

Meanwhile, frustrations, anxiety, and paranoia led Cantor to suffer more mental breakdowns, and he was hospitalized in 1899. Then tragedy struck: his young son suddenly died at the end of that year. Cantor was hospitalized again in 1903 and repeatedly thereafter. But at least some mathematicians developed his ideas. Poincaré complained: "Those mathematicians did not think that they were mistaken: they believed that they had the right to do what they did."[19] In Göttingen, David Hilbert prominently defended Cantor's theory in biblical terms: "Nobody can expel us from the paradise that Cantor has created for us."[20]

At Cambridge, the philosopher Ludwig Wittgenstein mocked the followers of Cantor. "I would not dream of trying to drive anyone out of this paradise," Wittgenstein said. "I would try to do something quite different: I would try to show you that it is not paradise—so that you'll leave of your own accord. I would say, 'you're welcome to this; just look about you.'"[21] He explained: "I am not saying transfinite propositions are false, but that the wrong pictures go with them. And when you see this the result may be that you lose interest." He argued that the notion or word *infinity* was being misused in this context, that Cantor's propositions might be correct but that their meaning, as commonly presented, was confused.[22] And he further commented:

Imagine infinite numbers used in a fairy tale. The dwarves have piled up as many gold pieces as there are cardinal numbers—etc. What can occur in this fairy tale must surely make sense.—

Imagine set theory's having been invented by a satirist as a kind of parody of mathematics.—Later a reasonable meaning was seen in it and it was incorporated into mathematics. (For if one person can see it as a paradise of mathematicians, why should not another see it as a joke?) The question is: even as a joke isn't it evidently mathematics?—And why is it evidently mathematics?—Because it is a game with signs according to rules? But isn't it evident that there are concepts formed here—even if we are not clear about their application? But how is it possible to have a concept and not be clear about its application?[23]

Wittgenstein was not a mathematician, but he worked on logic and the philosophy of mathematics for many years. He became convinced that mathematical propositions are not true, because they do not essentially refer to the real world. Against Platonism, he claimed that mathematical objects do not exist independent of our minds. He argued that various kinds of mathematics are invented, for various reasons, and that although axioms can be articulated and used to derive or prove propositions, the proof paths do not preexist our construction of them. Wittgenstein denied that infinity is a number because he conceived mathematical extensions as necessarily finite sequences of symbols, so an infinite mathematical extension was self-contradictory.

Most mathematicians who cared enough to actually read or consider Wittgenstein's views on mathematics thoroughly disagreed with him. I think that a major problem with Wittgenstein's views is that he argued that mathematics was invented *in its entirety*. He even came to regard numbers and numerals as being the same thing.

Let's assume, instead, that some numbers are not fictions, but concepts based on physical relations that exist independent of humans. I don't think that anybody invented the number or quantity 4; it represents real physical relations. We should then ask: do all numbers exist independent of our minds?

The number $\pi = 3.14159265\ldots$ has been calculated to trillions of decimal places. Does it preexist such calculations? Let's call its first digit, 3, "position 0," and its next digit, 1, "position 1," and 4 "position 2," and so forth. Mathematicians have pored through the many digits of π looking for any peculiar patterns. And they have found some. At position 768 there occur six nines in succession: 999999. The next block of six identical numbers is at position 193,034, remarkably, another set of 9s. Position 16,470 has the number 16,470. Position 44,899 has the number 44,899. Position 79,873,884 has the number 79,873,884. Much later, at position 17,387,594,880, there occurs, for the first time in π, the sequence 0123456789. Such coincidences prompt questions: where does your birthday show up in π? Where does a given sentence, coded in numbers, show up? The sequence of numbers is said to be *infinite*. Does the Bible show up somewhere in π? Or what about your entire chain of DNA: is it in there? Does it contain your thoughts? Consider the more than a trillion digits of π that computers have calculated: did that particular sequence of numbers preexist, somewhere, before we calculated them? Where?

So are numbers invented or discovered? Suppose that we identify an alien civilization, maybe on another planet: might we then communicate with them clearly by sending them some numbers, some equations? Would they promptly recognize π and know what we're talking about? Would they agree with −1 × −1 = +1? People who think of math as a "universal language" will answer such questions in the affirmative. Or, if someone claims instead that "numbers are models," that "math consists of making useful models of things," then we might ask: "Is π a model? Can we eventually invent a better model of π and thus realize that our current π is flawed?" Previously, I quoted an interview in which a teacher of mathematics was asked: "Does the number pi exist apart from people? Would the little green man from Galaxy X-9 know about pi?" The teacher replied: "As one gets older, one is less and less inclined to trouble oneself about this kind of question."[24]

Instead of avoiding such questions, we can well consider some of the alternative philosophies of mathematics that can help to orient us:

NAIVE PHYSICALISM

This is not a philosophy, but it takes the place that a philosophy should replace. It is the tacit view of many students in school and of many adults who have seldom thought carefully about mathematics. It involves the vague idea that some physical objects are "perfect" figures, for example, that waves made by a pebble in a lake are really circles, that the orbits of planets are really ellipses, and so on. It also involves related false assumptions about some numbers: for example, that humans know of physical objects that actually have *pi* or *phi* in their dimensions, when really, no such objects have been identified; in every case their dimensions only resemble such numbers. Even most philosophers in antiquity seem to have known that this outlook was wrong, so it is fair to try to cure students of such beliefs.

PLATONISM

Mathematical objects exist, objectively, independent of our knowledge of them. Numbers, triangles, circles, infinite sets, and all the rest are all real entities with definite properties, some not yet known. They are immaterial, existing outside of physical space and time. They are unchangeable and eternal, they do not evolve, and we can investigate the relations among them. "According to Platonism, a mathematician is an empirical scientist like a geologist; he cannot invent anything because it is all there already. All he can do is discover."[25]

FORMALISM

Mathematical objects do not exist; there are no such things. Mathematics consists of symbols, formulas, axioms, definitions, theorems. We establish those formulas by conventions, but systematically, and they are not essentially about anything: they are just strings of ordered symbols. Formulas can gain meaning if we choose to apply them to particular contexts such as physics, but they are essentially independent of such contexts. Mathematics is not discovered: it is invented.

These are the three major outlooks that we have considered, but there are others. Several of the apparent paradoxes that we have discussed have to do with infinity: division by zero, noncutting lines stretched to infinity, infinitely small numbers. Accordingly, some mathematicians, a minority, have decided to renounce the concept of infinity as a notion alien to genuine mathematics. They believe or think that legitimate mathematics should be constructed from definite finite quantities. They dislike discussions about infinity as meaningless nonsense, misapprehensions, disposable and unnecessary, improper to use in proofs. This philosophy is called finitism:

FINITISM

True mathematical objects exist independent of our inventions, yet not all so-called mathematical entities are legitimate. In particular, entities that are allegedly infinite are disposable fictions. Instead, all mathematical objects must be constructed from the finite quantities, the integers, in a finite number of steps.

This view was advocated by Cantor's critic Leopold Kronecker, who in an 1886 lecture in Berlin famously claimed, "The dear God made the integer numbers, all else is the work of man."[26] He worked to replace infinite notions, as well as negative and imaginary numbers. But Kronecker's views remained unpopular because so much of mathematics involves infinities. A related but more general outlook also arose, by which the existence of mathematical objects is admitted only if they can be constructed:

CONSTRUCTIVISM

True mathematical objects exist independent of our inventions, yet not all so-called mathematical entities are legitimate. All mathematical objects must be constructed rather than inferred; for example, we cannot assert the existence of something in math simply because its opposite is false. Constructivism restricts the kinds of proofs that can be used in math.

Some mathematicians, such as David Hilbert and Hermann Weyl, were very critical of the constructivist program, and they were pessimistic about the claim that much of mathematics could be developed by constructive methods. Yet in 1967 Errett Bishop showed that many theorems of real analysis could be established constructively.[27] Another prominent constructivist was Andrey Andreevich Markov. More particular versions of constructivism have been pursued by several mathematicians. For example, L. E. J. Brower became skeptical of the validity of some of the principles of classical logic and developed "Intuitionism," the outlook that mathematics is the constructive mental activity of humans, which can be decided or established in finite numbers of steps.[28]

By one mathematician's estimate, the beliefs of mathematicians around 1970 were distributed in the following proportions: Platonists 65 percent, formalists 30 percent, and constructivists 5 percent.[29] In contradistinction, other writers note that most mathematicians are not consistent in how they think about mathematics. According to Philip Davis and Reuben Hersh: "Most writers on the subject seem to agree that the typical working mathematician is a Platonist on weekdays and a formalist on Sundays. That is, when he is doing mathematics he is convinced that he is dealing with an objective reality whose properties he is attempting to determine. But then, when challenged to give a philosophical account of this reality, he finds it easiest to pretend that he does not believe in it after all."[30]

Various writers have proposed compromises on how to view mathematics. For example, Davis and Hersh argue that mathematics is a human invention, one of the humanities, in which we investigate true facts about imaginary objects. It's an interesting idea, but I disagree. I think that some parts of mathematics are discovered, while others are invented.

Numbers such as 5 and 17 are not arbitrary human inventions: they arise as representations of physical quantities and patterns that exist prior to or independent of humans. Basic addition is also rooted in experience. I think that large parts of mathematics, the most fundamental and important parts, were by no means invented. I don't think anyone invented 2 + 2 = 4, aside from the matter of numerals and notation. Instead, 2 + 2 = 4 is a symbolic proposition that represents a real and recurring relation among physical things, many of which existed prior to human beings and most of which exist independent of us. If all humans someday cease to exist, it would still be true that 2 + 2 = 4; that is, the physical relations it codifies would continue to be true. Such a relation exists at least in the physical world, and likewise, countless many other mathematical relations were really *discovered* rather than invented.

Yet other parts of mathematics are invented, such as the rules for operating with negative numbers. People establish them by convention in order to generate a more general and encompassing mathematics. I think that imaginary numbers, for example, were invented to solve some rather meaningless problems. Afterward, some individuals managed to give them meaning by devising useful applications for them in geometry and physics. Would an alien civilization be able to solve problems that we solve, but without using imaginary numbers? Yes.

My aim in briefly discussing philosophies of mathematics is to convey issues that arise among certain viewpoints. The question is: what are the consequences of thinking about mathematics in one or another way?

The Platonist view gives advantages and disadvantages. One advantage is that mathematics overall might appear more attractive, for being universal and eternal, for being a key to ultimate truth. Students might consider pursuing mathematics, then, because it transcends the transient and merely personal aspects of human life. As we have seen, many mathematicians were driven by such feelings, so they are certainly fruitful. On the other hand, one disadvantage is that students might think that most of mathematics has already been discovered—by analogy to geography, that the main continents and oceans and the highest mountains have already been mapped, missing only details here and there. Students who adopt the Platonist outlook might also think that they cannot *invent* things in mathematics, and they will not try to invent new numbers or alternative deviant solutions to old problems.

By contrast, a student who develops a formalist outlook might instead think that mathematics is like a game, with some occasional freedom to change the rules. Then the likely attraction might be less, because that student will be less inclined to believe that mathematics deals with eternal truths. Still, some such students might be inspired to take a creative critical approach, to be more impudent in their judgment of proofs and assumptions. That's the sort of attitude that led eccentric individuals such as Oliver Heaviside and J. Willard Gibbs to create vector algebra, by positing that the square of an imaginary number is positive, and likewise led Abraham Robinson to create nonstandard analysis.

Next, the philosophy of constructivism has the advantage that someone who pursues it may be motivated to bridge branches or mathematics that were considered incommensurable. But it has the disadvantage that if everyone were a constructivist, then we would not have the high rates of progress facilitated by using traditional logical assumptions (such as the old principle that if a proposition is not false then it *must* be true, and vice versa).

In any case, I actually dislike "-isms." But as we have seen, philosophies are connected to creativity, and such views have contributed immensely to the growth of mathematics. But still, I should specify my own opinion, and just to match the others, it might as well be an -ism:

PLURALISM

Some parts of mathematics represent physical patterns that exist independently of our imagination. Other parts of mathematics are free inventions that resemble rules and conventions of languages and architecture. Multiple mathematical systems are not all equivalent. By developing and using concepts and rules we construct new mathematical propositions that can be surprising, beautiful, and useful.

12

THE CULT OF PYTHAGORAS

There was a serpent in the philosophic paradise, and his name was
Pythagoras.
    —Bertrand Russell, "How to Read and Understand History"

Mythology deals with gods and heroes, tales that are passed down especially in popular oral traditions. We began with Pythagoras, in a time
when religion and science mixed. I don't know if he really contributed anything to mathematics, but he became portrayed in the form of classic myths:
a wise demigod who started a Golden Age, a hero who solved problems and
knew the secret of immortality. In time, stories about him increasingly included science and mathematics, and teachers learned to ignore his older
tales about gods, sacrifices, and magical powers.

But still, other elements from the mythical tradition continued to spread.
We should then return to the question of whether Pythagoras proved the
hypotenuse theorem. Whether this claim was made by Galileo in a fictional
dialogue, or whether it is voiced nowadays by a schoolteacher or painstakingly
elaborated by qualified professors of ancient Greek mathematics, my
impression is that all such claims suffer from a traditional urge to credit
Pythagoras, to heap fame upon fame. Hence an imagined past shines like
a light that might orient us. Against my impression, a common defense is
to argue that whereas indeed there is no evidence that someone in fact *did
discover* this or that, there is no evidence either that he *did not,* and therefore
it is quite possible that he did. Humbug. Possibilities properly do not exist in
the past: either events happened, or they did not; our historical conjectures
are not "possible," they are merely conceivable, products of the imagination,
plausible fictions. They are welcome and useful, but they are scarcely history.

A popular book relates these allegedly historical claims:

> Pythagoras developed the idea of numerical logic and was respon-
> sible for the first golden age of mathematics. . . . After twenty years
> of travel Pythagoras had assimilated all the mathematical rules in
> the known world. . . . Pythagoras had uncovered for the first time the
> mathematical rule that governs a physical phenomenon and demon-
> strated that there was a fundamental relationship between math-
> ematics and science. . . . Pythagoras died confident in the knowledge
> that his theorem, which was true in 500 B.C., would remain true for
> eternity. . . . Pythagoras constructed a proof that shows that every
> possible right-angled triangle will obey his theorem. For Pythagoras
> the concept of mathematical proof was sacred, and it was proof that
> enabled the Brotherhood to discover so much.[1]

I find it astonishing that despite their familiarity and currency, all of these claims are fiction. Ancient peoples used to explain away puzzling phenomena by telling tales about gods, rather than with science. Likewise, throughout the centuries many writers have tried to explain the past by telling tales about Pythagoras. Instead, we can write accounts that piece together mosaics of extant evidence without masking guesswork as certainties.

Historian David Fowler has worked to reconstruct the state of Greek mathematics during the early years of Plato's Academy. To do so, he rightly begins by focusing on chronology: by first setting aside most of the later sources, the ones that, for example, attribute achievements to Pythagoras. Following Walter Burkert, Fowler explains that the habit of giving credit to Pythagoras comes not from early Greek mathematics, but from later educational tradition. Fowler gives an example from a second-century text that required students to work on the grammar of a sentence about Pythagoras, the philosopher. Subsequently, the image of Pythagoras continued to evolve in the minds of teachers and students. Fowler carried out an amusing survey, asking people to complete the sentence: "Pythagoras the _____ was born in Samos and later went to Croton." From about 190 replies, "40% said mathematician or some variant (geometer, mystic geometer, triangle theorist . . .), 28% said philosopher or some similar variant, 12% philosopher-mathematician, and the rest a very mixed bag of activities—number freak, bean-hater, vegetarian, polymath, new ager."[2]

Having read the first chapters of the present book, readers who have not studied the many accounts of Pythagoras's alleged achievements might now imagine that it is fairly obvious that such stories are just legends and therefore do not require critical elaboration. But actually, those of us who argue that they are legends constitute only a tiny minority, against thousands to the contrary. In view of the lack of descriptive accounts I therefore have tried to specify and discuss every significant piece of early ancient evidence pertaining to the alleged connection between Pythagoras and mathematics. I do not expect that such arguments alone can succeed against notions acquired since childhood, but supplemented by other examples of historical speculations, I think that familiar impressions can change.

The cult of Pythagoras grew from the urge to exaggerate the past. I've read that Pythagoras was the first philosopher, the first to prove that the Earth is a sphere, that it moves; that he was supposedly the first mathematician, the first true geometer, the first to discover the hypotenuse rule, to discover and hide irrational numbers, to discover numbers in the harmonies of musical strings; that from Egypt he imported geometry and masonry into Italy; that he allegedly surmised gravity in the heavens and relativity in the cosmos; that he invented a liberal and political education and overthrew despotic regimes; that he greatly influenced Plato, Jesus, and the druids; that he predicted earthquakes and stopped violent storms; that he stilled the waters with a word; that he knew the secrets of alchemy, astrology, and eugenics; that he was divine, the son of the Sun god Apollo, and was born several times, spoke with gods and spirits, and traveled to the underworld; that he knew the secret of happiness, healed the sick, raised the dead, and convinced an ox not to eat beans. As Porphyry said seventeen centuries ago: "Of Pythagoras many other more wonderful and divine things are persistently and unanimously related, so that we have no hesitation in saying never was more attributed to any man, nor was any more eminent."[3] A video documentary likewise claims: "Even though he is a virtually forgotten hero in today's society, Pythagoras contributed to each and every one of our lives."[4]

We wish that at least some of these stories were true, but lacking evidence, I don't believe any of them. The attributions of great feats to Pythagoras come from the urge to read between the lines, to give more meaning to the blank empty spaces around ink on a page, to fill the gaps, to mix fact and speculation. Whoever Pythagoras really was, his image evolved through the fibs of his fans and the cumulative distortions of countless careless writers. His portrayals evolved so much that in time they exerted far greater influence than the actual man. Therefore, we should not just silence such myths, for they influenced history: they should be heard, appreciated, and criticized.

At the start of this book, I noted that by being careful with sources, we can replace historical myths with accounts that are better and true. So what might we say instead of the usual: "Pythagoras of Samos was the first person to prove the Pythagorean theorem"? We might say:

> Because we care about *proof,* we admit that we have no proof of who first proved the hypotenuse rule; yet it is bizarre that for centuries many teachers gifted the credit to an ancient religious leader, Pythagoras, whose fame grew from writers' speculations.

I really think that this is a more interesting story; it highlights the importance of proof but acknowledges our fallibility.

As a whetstone serves to sharpen knives, in 1557 Robert Recorde characterized arithmetic as a means to sharpen our minds: "The Whetstone of Witte." Even if we do not regard Pythagoras as the author of mathematical proofs, I believe that we should all still study Pythagoras, not to memorize his alleged achievements but to sharpen our skepticism. The aim would not be to distrust everything Pythagorean but to analyze historical claims against evidence.

I'm increasingly unsatisfied by grand vague sentences such as "For better or for worse, the doorway of the Pythagorean Academy would eventually open, leading into the modern world."[5] Or in the words of Arthur Koestler, "The Pythagorean concept of harnessing science to the contemplation of the eternal, entered, via Plato and Aristotle, into the spirit of Christianity and became a decisive factor in the making of the Western world."[6] Instead of generalities, we should be specific. By avoiding the charming language of grand exaggerations, one surrenders spellbinding literary tricks, but I've tried to use words that are fair and transparent. Still, in the cozy blur of illusions, clarity is an unwelcome guest. So I hesitated: to avoid fiction is to leave the gushing currents that move readers to turn page after page with the intrigue of a dream. Would this book be as appealing as the novels that crowd the front aisles of bookstores? I remembered George Orwell's words: "It is bound to be a failure, every book is a failure."[7] Meticulous attention to details and primary sources might seem invisible. Hyperbole tastes sweet, but skepticism bitter. Critical writings kindle readers' doubts, like wildfire that quickly turns against a skeptical text itself.

Still, without exaggerations, the stories of mathematics are surprising. It seemed impossible that the elements of arithmetic might change. It seemed inconceivable that between the positive numbers and zero there might be room for other numbers—but some individuals dared to propose that there might well fit infinitely many strange numbers in that barely imaginable gap. And geometers believed, they knew, that given a straight line, on an outside point there could exist only one line parallel to the first. But that changed, partly thanks to the efforts of Lobachevsky, a lousy student who became a teacher. In 1843, an alcoholic mathematician with marriage problems, Hamilton, managed to break some rules of algebra by proposing new imaginary numbers. Later, a shy, unemployed high school dropout who lived with his parents much too long, Oliver Heaviside, managed to formulate a new algebra of vectors, by breaking the usual rules of multiplication. These are stories about change—where the apparently impossible became plausible or true.

Such amazing factual stories should accompany tales that used to be told as certain, say, that Pythagoras was a great mathematician. Now, do his many alleged achievements in this book exhaust his fame? Not at all! Remember what Iamblichus wrote: "But all the discoveries were of that man, for so they refer to Pythagoras, and do not call him by his name."[8] Among his further alleged achievements are also the following. Diogenes Laertius said that Aristoxenus the musician claimed that Pythagoras was "the first person who introduced weights and measures among the Greeks."[9] I've repeatedly also read that "he invented the multiplication table, called after him the Abacus Pythagoricus."[10] Supposedly he was "the father of logic."[11] Allegedly, he also "conjectured that the milky way and the nebulae consisted of innumerable small stars."[12] And Carl Sagan echoed a claim by Laertius, that Pythagoras was the first person to give the name "cosmos" to the universe.[13] Why not just attribute the rise of all science to *that man*? Well, some writers have. In a bestselling book, which I recommend, Koestler voices exuberant claims, which I reject: "Pythagoras of Samos was both the founder of a new religious philosophy, and the founder of Science, as the word is understood today."[14] Part of this tendency is just the traditional urge to attribute the origins of Western civilization to the ancient Greeks. Koestler insisted: "Plato and Aristotle, Euclid and Archimedes, are landmarks on the road; but Pythagoras stands at the point of departure, where it is decided which direction the road will take. Before that decision, the future orientation of Greco-European civilization was still undecided: it may have taken the direction of the Chinese, or Indian, or pre-Columbian cultures, all of which were still equally unshaped and undecided at the time of the great sixth-century dawn."[15]

Elsewhere I read that in 500 BCE, Pythagoras was the first of the Greeks who recognized ethics as a social force.[16] And writers have made various claims about Pythagoras's religion. That he discovered that all souls are eternal.[17] That he believed in many gods and demigods. That he sacrificed to them. That he never sacrificed to them. Or that he toiled for the gods but concluded at last "that there is a nature and spontaneousness in all things, and that the gods have no care for men."[18] Or that he believed in the one God that contains all things.[19] That he admired and imitated Jewish doctrines.[20] That in Egypt he learned the true name of God, Jehovah.[21] That he believed in the Holy Trinity.[22] That he did *not* believe in the transmigration of souls.[23] That he was an exorcist who cast out demons.[24] That he was a pantheist.[25] That he was secretly an atheist.[26] That "he brought from India the wisdom of the Buddha, and translated it into Greek thought."[27] So wait, what was he? Shaman, polytheist, pantheist, Jewish, Christian, Buddhist, or atheist?

Likewise, writers have portrayed Pythagoras as a great politician. Some scholars portray him as a political exile and revolutionary. Others argue that he organized a political aristocracy founded on popular ignorance.[28] Others say that he instituted a kind of communism.[29] One writer claims that Pythagoras was a socialist.[30] Another professor argued instead: "The first great exponent of democratic thought was Pythagoras, a citizen of Samos."[31]

In 1904, a psychology textbook portrayed Pythagoras as the man who proved the hypotenuse theorem and sacrificed to heaven, from which a wealth of discoveries followed: "For through him mankind has been able to measure the earth and weigh the mountains."[32] It claimed that Pythagoras "precipitated a crisis in human affairs—humanity could never be what it was before. He had discovered mathematical reasoning, thereby changing the course of history." This alleged breakthrough then supposedly led to the application of reasoning to politics, architecture, sculpture, painting, literature, and law, so that symmetry and consistency entered into human affairs: "Pythagoras having, then, set in motion the course of events which finally culminated in the Roman empire, well exemplifies the saying that truth is stranger than fiction."[33]

I don't think that Pythagoras revolutionized civilization. But truth is indeed stranger than fiction insofar as the historical past becomes a stranger, whereas historical fictions are quite familiar. Elsewhere I find yet another imaginative hyperbole: "Pythagoras is the founder of European culture in the Western Mediterranean sphere."[34] Thus it would seem that *it was he* who founded science and math, that he was the first great proponent of democracy, that he launched the rationality that led to the Holy Roman Empire, and that he founded European culture. I don't think so.

Without fair warning, any reader who innocently encounters a brief mention of Pythagoras in a textbook on math, science, or music might automatically accept it as a valid historical marker. But by gaining a sense of the extent to which Pythagoras has been conjured in many fields, I hope that readers will realize that most such mentions of Pythagoras are less than history. He has been hailed in philosophy, religions, ethics, feminism, vegetarianism, animal rights, music, astronomy, arithmetic, number theory, numismatics, geometry, geography, geology, alchemy, medicine, physiognomy, eugenics, communism, socialism, democracy, architecture, freemasonry, druidism, fortune telling, and magic. There's something wrong here!

In 1968, sociologist Robert K. Merton published an article analyzing peculiarities in the allocation of credit among scientists. He analyzed interviews with winners of the Nobel Prize, who repeatedly commented that they get too much credit for their contributions, while other scientists get disproportionately little credit for comparable contributions. One Nobel laureate commented: "The world is peculiar in this matter of how it gives credit. It tends to give credit to [already] famous people."[35] Merton analyzed the complex pattern of the misallocation of credit in science and called it "the Matthew effect," referring to two passages in the New Testament, in the Gospel of Matthew (King James Version):

> For whosoever hath, to him shall be given, and he shall have more abundance: but whosoever hath not, from him shall be taken away even that he hath. (13:12)

> For unto every one that hath shall be given, and he shall have abundance: but from him that hath not shall be taken away even that which he hath. (25:29)

The so-called Matthew effect is so pervasive that it even involved ironies in the case of Robert Merton himself. When five years later he reissued his original article, in 1973, he noted that he should have shared authorship for that article with Harriet Zuckerman, because she had carried out the extensive interviews with Nobel laureates.[36] Moreover, Merton acknowledged that there were objections to calling this "the Matthew effect," because this terminology seems to credit only Matthew, whereas similar passages appear in the gospels of Mark and Luke. And because it seemed doubtful that Matthew really authored the "Gospel of Matthew." And because the three evangelists were presumably quoting Jesus.[37]

Furthermore, Merton noted that Marinus de Jonge, professor of theology at the University of Leyden, had further explained that "it is highly likely that [Jesus] took over a general saying, current in Jewish (and/or Hellenistic) Wisdom circle."[38] And now the qualifier "Hellenistic" gives us again the tempting opportunity, of course, to give the credit to Pythagoras.

The "Matthew effect" greatly affects stories about Pythagoras, Einstein, and others. The case of Pythagoras seems to me the most astonishing both because so much has been ascribed to him and because there is so little evidence that he really contributed anything to science or mathematics at all. His main advantage over others is time. Given thousands of years, the allocations of credit to his name grew spectacularly.

There's a pervasive habit to give credit where it is not due, to feign a modicum of history. This habit is so strong that it becomes possible to make predictions of what people might claim in the future or the past. In 2009, a news article tells me, without evidence, that Pythagoras advanced the possibility of space travel.[39] Really? Writers on the Internet now claim that he was the first proponent of the physics of string theory. Predictable. When I first began this project, for example, I did not know whether there were in fact connections between Pythagoras and the golden ratio, but having seen various free-wheeling attributions, I suspected that there must be some writers who had made the connection. Researching, I found that such stories indeed had developed. Likewise, while researching a previous book, Science Secrets, I also suspected alleged connections between Pythagoras and alchemy, and I was stunned to see the extent to which such stories had developed.

There is also a tendency to give credit to distant authorities rather than to recent peers, as if the greatness of a breakthrough would be diminished by association with living persons. This tendency was strong in Newton, for example, as he recoiled from granting credit to his peers such as Leibniz and Robert Hooke: "Though Newton could speak freely in praise of Moses and Toth, Thales, Pythagoras, Prometheus, and Chiron the Centaur, he only rarely had good words to say about either living scientists or his immediate predecessors—."[40] In my previous book I show how Newton ascribed his theory of gravity to Pythagoras, and how Kepler surmised that the Pythagoreans knew of the five solid figures in the heavens, and that Charles Lyell speculated, on the basis of a mere poem, that Pythagoras might have founded the geological theories of catastrophism and uniformitarianism. So we should recognize that this habit can affect us too. It need not manifest itself in attributions to Pythagoras in particular, but more generally, in the reflex to feign knowledge of the past.

Questions of priority and authorship are a thin slice of history, but they are symptomatic of the urge to invent history. False stories arise by a cumulative process of taking different but similar claims as equal, disregarding the original details of a statement to make it say something else. Consider how some astronomical breakthroughs became attributed to Pythagoras. A fair account by Aristotle seems vague, so almost unconsciously, writers modify it. Around 350 BCE, Aristotle argued that some individuals who called themselves "Pythagoreans" believed that the Earth moves, and the Sun too, around a "central fire." That ancient account then evolved by a series of alterations. Some writers generalized that the Pythagoreans believed that the Earth and the planets move in circles *around the Sun.* Next, others wrote that Pythagoras *himself* taught his followers that the Earth and the planets circle the Sun. Later writers even claimed that Pythagoras knew the inverse square law of gravity.[41] And likewise, that Pythagoras "placed the sun in the centre and all the planets moving in elliptical orbits around it."[42]

I don't think that social and economic factors forced each writer to distort the previous accounts. It would be a good story, and I'd be happy to tell it, but I have no certainty about any causal conjunctions of circumstances that moved individuals to distort the past. Instead, it seems to be a recurring carelessness, a lack of focus on the printed text, a reluctance to literally echo what it says, in favor of apprehending instead a meaning that more closely matches our concerns. We see this tendency at home, on television, in newspapers, even at universities. There are many causes for it; one is that prior to reading or listening, we already hold in our minds certain schemes into which we try to fit what we learn.

It is not just nitpicking to scratch such stories. There are formidable forces at work. First they seem negligible, as adding slight literary ornaments seems fine, making no significant difference, just like adding infinitesimals. But the cumulative effect of such distortions, over time, is toxic because it crowds out the historical or documentary realities. These are the forces that gave Pythagoras most of his powers, and these are also the forces that buried him.

The great writer Arthur Koestler tried to shrug off the bothersome notion that maybe beneath the tales of Pythagoras there was not much substance. Koestler chose to assume that only a powerful seminal insight could exert lasting and pervasive influence on humanity. He argued: "Myths grow like crystals, according to their own, recurrent pattern; but there must be a suitable core to start their growth. Mediocrities or cranks have no myth-generating power; they may create a fashion, but it soon peters out. Yet the Pythagorean vision of the world was so enduring, that it still permeates our thinking, even our vocabulary."[43] Maybe, but I disagree that cranks have no myth-generating power. Consider the influence of Nostradamus, the alleged prophet. A search on the Internet for Nostradamus right now gives me 17.9 million hits in Google. A search for "Pythagoras" gives much fewer, 6.2 million. So who had the more powerful vision? Who sells more books and magazines? Who has more often been the focus of television shows? Pythagoras might seem much more reasonable, but that's partly because we focus mainly on the scientific and mathematical associations with his name. But there is a world of pseudoscience and mysticism where Pythagoras's alleged achievements were also supposedly epoch-making.

For example, there are neglected pagan doctrines that, I have argued, should be taken into account when we analyze the Catholic proceedings against "the Pythagorean doctrine" that the Earth moves, in the days of Galileo.[44] Over the centuries, the so-called Pythagoreans made various claims: that the Sun, the Moon, and the stars are gods; that human souls come from the Milky Way; that stars are worlds; that souls go to the Sun; that Pythagoras was a demigod; that Pythagoras visited the underworld; that he spoke to gods and demons, performed miracles and magic. But furthermore, such mystical allegations continued to multiply, centuries later. The esoteric medium and plagiarist Madame Blavatsky claimed that Pythagoras "laid the origin of the differentiated cosmic Matter in the base of the Triangle."[45] Reputedly, Pythagoras was a master of divination: numerology, hydromancy, geomancy, and onomancy—that is, by deciphering the future on the basis of numbers, water, bits of dirt, or words, respectively.[46] Voltaire quipped that *everybody knows* that Pythagoras knew the language of animals and plants.[47] Allegedly he practiced "mesmerism," subduing wild animals with his voice and touch, and also he could "write on the Moon" by smearing blood on a mirror and orienting it to project the image onto the Moon.[48] And in the 1930s, the president of a spiritualist society in New York claimed to report dictations clairaudiently received from the spirit of Pythagoras: allegedly he said that the Earth originated from two electrons united by "the simple law of addition by attraction—Love." At least this alleged voice of Pythagoras also said: "In the many translations and the over-much editing of my writings there has ensued a wholesale slaughter of both false and true."[49]

Some historians have admitted that some of the earliest extant accounts of Pythagoras (such as by Heraclitus, Isocrates, and Xenophanes) seem to portray him as a charlatan. So it seems conceivable that the most famous name in the history of mathematics was actually what many mathematicians and scientists despise: a crank, a crackpot, a charlatan. Stories can grow regardless of the original kernel of truth or fool's gold, partly because the forces that impel them are not seminal, as Koestler implied. Many forces are external: the audience decides, the market, the popular imagination.

Moreover, we look to history to tell us not only what happened but *why*. We tend to be unsatisfied with a documentary accounting that chronologically lists historical evidence. We want to know the causes, though those are often inaccessible. Did Pythagoras eat beans? I really don't know, yet the ancient claim that he forbade his disciples to eat beans led various writers to eagerly guess *why*. Because they cause gas, because they upset the stomach, because they cause nightmares, or because they resemble parts of the human body or "are like the gates of hell (for they are the only plants without parts)," or because they dry up other plants, or because they represent a universal nature, or because they were used in government elections.[50] Or because beans dull the senses, or cause sleeplessness, "or because the souls of the dead are contained in a bean."[51] Or for many other reasons, including that some people around the Mediterranean are seriously allergic to fava beans.

One reason why so many stories evolved about Pythagoras is because the superabundance of tales about him by 450 CE enabled writers to pick and choose whichever components seem preferable. Likewise, myths about Einstein have grown from bits of apparent evidence. I've used Pythagoras and Einstein as the endpoints of this book and of *Science Secrets* because while in the case of Pythagoras we receive him as a legend, in the case of Einstein we can closely analyze the gradual processes whereby his image has been construed into mythical forms. Pythagoras was a religious leader who eventually became misrepresented as a great mathematician and astronomer. Einstein was a physicist who eventually became misrepresented as a mathematician and a religious saint.

Einstein, Pythagoras, and history on the whole are used as a kind of currency. They add value to statements, true or apparent; they help to guide a compass that tells us what's right. The name "Pythagoras" still rings bells. The bestselling book *The Secret* says that he knew the secret of happiness, like other innovators: Shakespeare, Beethoven, Leonardo, Newton.[52] By heaping fame onto Pythagoras, writers have converted him into a brand of value.

By association, ideas acquire a noble past. In 1789, prior to the French Revolution, Jean-Jacques Barthélemy published a fictional but inspirational dialogue in which he portrayed the ancient glory of "The Institute of Pythagoras," which aspired to order, perfection, and equality. Barthélemy cited ancient texts to portray Pythagoras as a divine role model. Echoing the old claim that he established peace and harmony within and among the Greek nations, Barthélemy also claimed that Pythagoras taught how to attain purity and health, how to rightly share a community of goods, and how to live with justice and equality. He noted that "the peoples realized that a god had appeared on Earth to deliver them from the evils that afflicted them."[53] And he proclaimed: "That which ensured his glory, that which he conceived as a grand project: is that of a congregation, which always subsists and always as a repository of sciences and mores, being the organ of the truth and of virtue, when men will be in a state of understanding the one and practicing the other."[54]

Barthélemy, like most writers, dismissed the legendary nonsense: "Surely you do not believe that Pythagoras advanced the absurdities that are attributed to him?"[55] Yet Barthélemy accepted the useful heroic stories and dismissed whatever he didn't like. His account became very popular among intellectual revolutionaries who sought a model for a moral and political brotherhood. Accordingly, in 1799, Sylvain Maréchal, a French political theorist and Freemason, published a six-volume treatise titled *Voyages of Pythagoras*.[56] Maréchal portrayed Pythagoras as the model intellectual-turned-revolutionary, one who founded a brotherhood that advocated ideals of common ownership toward a republic of equals. The books by Barthélemy and Maréchal were issued widely in French, German, and Russian. The ideals propagated, for example, as a Russian journal advocated Maréchal's 150 "Rules of Pythagoras."[57] Various Russian youths organized activist groups, "circles," inspired by the alleged rules and laws of Pythagoras, in the interest of equality.[58] Fyodor Glinka, a Russian poet and soldier, was apparently influenced by Barthélemy's account of the "Institute of Pythagoras" and became a leading participant in the secret political society the Union of Salvation, founded in 1816.[59] Its strict procedures for initiation and promotion emulated the system of the Freemasons. Meanwhile, in 1923, Lenin, the paralyzed leader of the Soviet Union, denounced the Pythagorean number theory and the connection of the rudiments of science with fantasy and mythology.[60] Regardless, former members of the secretive Union of Salvation participated in the bloody military revolt of December 1825.

Furthermore, the image of Pythagoras was sometimes used by opportunists who feigned nobility. Some individuals used Pythagorean lore to take advantage of others. Around 430 BCE, Herodotus told the story of a slave of Pythagoras, Salmoxis, who after being freed from slavery managed to gain wealth. To take advantage of poor and ignorant people, Salmoxis constructed a hall where he entertained and fed local leaders

> and taught them that neither he nor his guests nor any of their descendants would ever die, but that they would go to a place where they would live forever and have all good things. While he was doing as I have said and teaching this doctrine, he was meanwhile making an underground chamber. When this was finished, he vanished from the sight of the Thracians, and went down into the underground chamber, where he lived for three years, while the Thracians wished him back and mourned him for dead; then in the fourth year he appeared to the Thracians, and thus they came to believe what Salmoxis had told them. Such is the Greek story about him.[61]

Likewise, in 56 BCE Cicero criticized the opportunistic use of the lore of Pythagoras. Cicero denounced the politician Vatinius for calling himself "a Pythagorean, and to put forth the name of a most learned man as a screen to hide your own savage and barbarian habits,—what depravity of intellect possessed you, what excessive frenzy seized on you, and made you, when you had begun your unheard-of and impious sacrifices, accustomed as you are to seek to evoke the spirits of the shades below, and to appease the spirits of the dead with the entrails of murdered boys."[62]

These last examples are significant because they highlight a deceptive strain of Pythagorean trends. Since antiquity, the lore of Pythagoras often involved a sort of pretension. Supposedly Pythagoras could clearly remember everything, even about his past lives, an ability gifted by the god Mercury.[63] The fair ambition to know the past unfortunately devolves into the whim to invent it, to pretend to know. "Pythagoras" is an emblem of our impatient urge to know, a mask for ignorance, a bluff in the teaching game, a crutch. It's a blank slate, on which to pretend to write the past. Whether your field is mathematics, philosophy, music, medicine, numismatics, geography, geology, astronomy, religion, politics, or magic, there are claims that Pythagoras made major great contributions to it. Whether you believe in one god or many, whether you believe in the Holy Trinity or not, whether you're a vegetarian or sacrifice animals, whether you follow mainly Jesus or the Buddha, there is a Pythagoras for you. Most tales of Pythagoras are symptoms of our unwillingness to confront uncertainty, to plainly admit: I don't know what happened.

Like other philosophers fascinated by mathematics, science, and history, Bertrand Russell was very impressed by Pythagoras. Russell commented: "I do not know of any other man who has been as influential as he was in the sphere of thought."[64] And like other writers, relying mostly on what can charitably be described as secondary sources, Russell too succumbed to the urge to attribute to Pythagoras epoch-making achievements: that demonstrative deductive mathematics began with him, that he founded a school of mathematicians, and even that "the whole conception of an eternal world, revealed to the intellect but not to the senses, is derived from him. But for him, Christians would not have thought of Christ as the Word; but for him, theologians would not have sought logical *proofs* of God and immortality. But in him all this is still implicit."[65]

Yet Russell also expressed misgivings at the Pythagorean blend of reason and mysticism. He argued that Greek philosophy had a brilliant beginning but that its development was poisoned by Pythagoras, the "serpent" who imported elements of the Orphic religion, which did not aim to honestly understand the world, but to pursue a kind of spiritual intoxication:

> From that day to this, there has been thought to be something divine about muddleheadedness, provided it had the quality of spiritual intoxication; a wholly sober view of the world has been thought to show a limited and pedestrian mind. From Pythagoras this outlook descended to Plato, from Plato to Christian theologians, from them in a new form, to Rousseau and the romantics and the myriad of purveyors of nonsense who flourish wherever men and women are tired of the truth.[66]

Russell advocated science as a hopeful cure against submersion in old and new superstitions. It is encouraging to see that this lucid writer tried to shake off the spell of Pythagoras. As with Koestler, I'm very impressed by Russell's words, but again I can't wholly agree with him either. As I reject Koestler's willingness to credit Pythagoras with the seed of Western civilization, I also reject Russell's notion that Pythagoras was guilty, to be blamed for the reams of pretension that arose in his name. From an utter lack of evidence, we can hardly believe that Pythagoras founded science and deductive mathematics and admire him for that, yet I do not see any reason to condemn him as having inflicted an original poison into philosophy either. Most of the pleasant fictions and nonsense that developed over the centuries as "Pythagorean" are quite distinct from the ancient sources.

Still, there were regrettable pretensions and obfuscations. Reportedly, the Pythagoreans obstructed the revelation of true knowledge to just anyone, they opposed the disclosure of knowledge in any easily intelligible way, they opposed individual freedom, and hence the alleged authority of the Master prevailed unsupported by reason. And they excelled at the practice of giving credit where credit was not due. Such stories, usually omitted, deserve to be told so that we might recognize how often we behave like members of the cult.

The bad habits that distort Pythagoras also distort other figures and events in history. The case of Pythagoras shows how seemingly harmless little infidelities, when we paraphrase stories and documents, can bury the extant traces of the past. Rather than polish images of the dead, we should better polish the mirror of history. Some haunting ghosts fade, but we gain glimpses of what they hide.

NOTES

Introduction

1. Rhonda Byrne, *The Secret* (New York: Atria Books, 2006), 4; Carl Sagan, *Cosmos* (New York: Random House, 1980), 183; John Wood, *Choir Gaure, Vulgarly called Stonehenge* (London, 1747), 11–12.

2. Paul Lockhart, *A Mathematician's Lament* (New York: Bellevue Literary Press, 2009), 40, 45, 51, quotation at 83–84.

3. Robert Crease, "This Is Your Philosophy," *Physics World* (Apr. 2002).

4. John Allen Paulos, *Innumeracy: Mathematical Illiteracy and Its Consequences* (New York: Hill and Wang/Farrar, Straus and Giroux, 1988), 80.

1. Triangle Sacrifice to the Gods

1. Hippolytos, *Hippolyti Philosophumena*, chap. 2, in Hermann Diels, *Doxographi Graeci* (G. Reimer: Berlin, 1879), 555, also in *The First Philosophers of Greece,* ed. and trans. Arthur Fairbanks (New York: Charles Scribner's Sons, 1898), 153–54.

2. Since historians argue that Jesus Christ was born four years before the date previously believed, the expression BCE, meaning "before the common era," is preferable to BC partly because it avoids the bizarre chronological statement that Jesus Christ was born four years "before Christ."

3. On the Chinese contributions, see Frank J. Swetz and T. I. Kao, *Was Pythagoras Chinese?* (University Park: Pennsylvania State University Press/National Council of Mathematics Teachers, 1977). Contrary to widespread claims, no evidence has yet been found that the ancient Egyptians knew the theorem, not even for a triangle of side lengths 3, 4, and 5; see Richard J. Gillings, *Mathematics in the Time of the Pharaohs* (Cambridge, MA: MIT Press, 1972; reprint, New York: Dover, 1982).

4. Diogenes Laertius, "Life of Pythagoras," in *The Lives and Opinions of Eminent Philosophers* (ca. 225 CE), trans. C. D. Yonge (London: Henry G. Bohn, 1853), book 8, sec. 15.

5. Claudius Ælianus, *Varia Historia* (ca. 220 CE); "Wonders and Opinions of Pythagoras," in *Claudius Ælianus, His Various History,* trans. Thomas Stanley (London: Thomas Dring, 1665), book 4, sec. 17.

6. Diodorus Siculus, *Bibliotheca Historica* (ca. 50 BCE), in *Diodorus of Sicily in Twelve Volumes,* trans. C. H. Oldfather, vols. 4–8 (Cambridge, MA: Harvard University Press, 1989), book 10, sec. 3.

7. Iamblichi, *De Vita Pythagorica* (ca. 300 CE), reissued as Iamblichus, *On the Pythagorean Way of Life,* ed. and trans. John Dillon and Jackson Hershbell (Atlanta: Scholars Press, 1991), chap. 28, p. 154. See also p. 155, where the editors note that Porphyry stated that the river was the Caucasus, while Aelian stated that the river was the Casas, near Metapontium.

8. Ancient poem quoted in Iamblichus, *On the Pythagorean Way,* chap. 2, p. 35.

9. Albert Einstein, draft of an address of Apr. 1955, Einstein Archives, item 60-003.

10. Proclus, *Eis proton Eukleidou stoicheion biblon* (ca. 450), in *Procli Diadochi in Primum Euclidis Elementorum Librum Commentarii,* ed. Godofredi Friedlein (Lipsiae: B. G. Teubneri, 1873), 426 (trans. Martínez).

11. Walter Burkert, *Lore and Science in Ancient Pythagoreanism,* trans. Edwin Minar (Cambridge, MA: Harvard University Press, 1972), 409–12; Iamblichus, *De Communi Mathematica Scientia Liber,* ed. Nicolaus Festa (Lipsiae: Teubner, 1891), book 70, sec. 1.

12. Iamblichus, *On the Pythagorean Way,* chaps. 27, 32, 7, pp. 149, 215, 59, respectively.

13. Iamblichus, *On the Pythagorean Way,* chaps. 25–26, pp. 137–43, chap. 6, pp. 55–57.

14. Iamblichus, *On the Pythagorean Way,* chap. 12, p. 83.

15. Iamblichus, *On the Pythagorean Way,* chaps. 24, 13, 19, pp. 131, 85, 117, respectively.

16. Iamblichus, *On the Pythagorean Way,* chap. 2, p. 37.

17. Porphyry, *Adversus Christianos* (ca. 290? CE). Some fragments remain: see R. Joseph Hoffman, *Porphyry's* Against the Christians: *The Literary Remains* (Amherst: Prometheus Books, 1994).

18. Porphyry, *Life of Pythagoras* (ca. 300? CE), in *The Pythagorean Sourcebook and Library,* trans. Kenneth Sylvan Guthrie (New York: Platonist Press, 1919), rev. ed. David R. Fideler (Grand Rapids: Phanes Press, 1987), sec. 28, p. 128.

19. Porphyry, *Life of Pythagoras,* sec. 36, in Porphyriou, *Pythagorou Bios* (ca. 300? CE), in *Vie de Pythagore* (Paris: Les Belles Lettres, 1982), sec. 36 (trans. Martínez).

20. Diogenes Laertius, "Life of Pythagoras," book 8, sec. 11, in *Lives of Eminent*

Philosophers, ed. R. D. Hicks (Cambridge, MA: Harvard University Press, 1972) (trans. Martínez).

21. Leonid Zhmud, "Pythagoras as Mathematician," *Historia Mathematica* 16 (1989): 249–68, quotation at 257.

22. Diogenes also cited another source: "And Pamphila relates that he [Thales of Miletus], having learnt geometry from the Egyptians, was the first person to describe a right-angled triangle in a circle, and that he sacrificed an ox in honour of his discovery. But others, among whom is Apollodorus the calculator, say that it was Pythagoras who made this discovery." See Diogenes Laertius, *Lives and Opinions,* book 1, sec. 3.

23. Athenaeus, *Deipnosophistai* (ca. 200 CE), in *The Deipnosophists,* ed. and trans. Charles Burton Gulick (Cambridge, MA: Harvard University Press, 1928), book 10, chap. 13 (trans. Martínez).

24. Plutarch, ΟΤΙ ΟΥΔ΄ ΗΔΕΩΣ ΖΗΝ ΕΣΤΙΝ ΚΑΤ΄ΕΠΙΚΟΥΡΟΝ (That It Is Not Possible to Live Pleasurably by the Doctrine of Epicurus), (ca. 100 CE), in Plutarque, *Oeuvres Morales,* vol. 5, ed. and trans. into French by Victor Bétolaud (Paris: Hachette, 1870), sec. 11 (trans. Martínez).

25. In antiquity, multiple men had the name Apollodorus, even some famous individuals, including playwrights, writers, and philosophers. For example, there was an Apollodorus of Athens (ca. 180–120 BCE), a writer and mythographer who wrote a Greek history in verse that was later imitated by forgeries and confused by misattributions.

26. Vitruvius, *De Architectura Libri Decem* (ca. 15 BCE), ed. Valentin Rose (Lipsiae: B. G. Teubneri, 1909), book 9 (trans. Martínez).

27. Marcus Tullius Cicero, *De Natura Deorum* (ca. 45 BCE), book 3, sec. 88 (trans. Martínez).

28. Diogenes Laertius, "Life of Pythagoras," sec. 12.

29. Eudoxus, *Description of the Earth,* book 2 (ca. 340 BCE), quoted or paraphrased in Porphyry, *Life of Pythagoras,* sec. 7, p. 124.

30. Ovid, *Metamorphoses,* book 15, fable 2.

31. Diodorus Siculus, *Diodorus of Sicily in Twelve Volumes,* book 10, chap. 6.

32. Aristotle, writing about 150 years after Pythagoras, seems not to have specified whether Pythagoras was a vegetarian, but allegedly his self-proclaimed admirers observed a few dietary restrictions, such as not eating heart or womb. Aulus Gellius, in *Noctes Atticae* (ca. 170? CE), claimed that Plutarch, in his first book on Homer, claimed that Aristotle made this claim. See Aulus Gellius, *Attic Nights,* trans. William Beloe, 3 vols. (London: J. Johnson, 1795), vol. 1, chap. 11, p. 264.

33. For example, see the discussion in Christoph Riedweg, *Pythagoras: His Life, Teaching, and Influence,* trans. Steven Rendall (Ithaca and London: Cornell University Press, 2005), 68–69.

34. Quoted in Diogenes Laertius, "Life of Pythagoras," sec. 38 (trans. Martínez).

Diogenes inferred that Xenophanes was referring to Pythagoras, but we only have this brief excerpt.

35. Robert Crease, "Critical Point: Pythagoras," *Physics World* (Jan. 2006); see also Zhmud, "Pythagoras as Mathematician," 257. In *The Great Equations,* Crease seems to refer to Pythagoras as having proved the theorem (27, 33), but in the endnotes Crease rightly casts doubt on this traditional claim. See Robert Crease, *The Great Equations* (New York: W. W. Norton, 2008), 278.

36. Plutarch, "A Discourse Concerning Socrates's Daemon," in *Plutarch's Miscellanies and Essays. Comprising All His Works under the Title of "Morals,"* rev. William W. Goodwin, 6th ed. (Boston: Little, Brown, and Company, 1898), vol. 2, sec. 9, p. 388.

37. Plutarch, "How to Know a Flatterer from a Friend. To Antiochus Philopappus," in Goodwin, *Plutarch's Miscellanies and Essays,* vol. 2, sec. 32, p. 148.

38. The oldest fragment copy of a portion of the *Elements* is from roughly 100 CE, dated paleographically by Eric Turner; see David Fowler, *The Mathematics of Plato's Academy,* 2nd ed. (Oxford: Oxford University Press, 1999). Writers usually date the *Elements* to ca. 300 BCE because at around 460 CE Proclus claimed that Euclid lived during the reign of king Ptolemy I (323–283 BCE). However, no evidence from the preceding seven centuries specifies when Euclid completed the *Elements.* Historian Alexander R. Jones, in an unpublished manuscript, argues that Proclus made arbitrary conjectures and that the scant evidence (including extant and lost references to Euclid in the works of Apollonius and Archimedes) suggests that the *Elements* was completed later in the third century BCE.

39. Plato, *Republic* (ca. 375 BCE), book 10, 600b.

40. E. T. Bell, *Men of Mathematics* (New York: Simon & Schuster, 1937; reprint, New York: Simon and Schuster, 1962), 21.

41. Constance Reid, *The Search for E. T. Bell, also known as John Taine* (Washington, DC: Mathematical Association of America, 1993), 290.

42. Galileo Galilei, "Dialogo Primo," in *Dialogo, Doue ne i Congressi di Quattro Giornate si Discorre sopra i due Massimi Sistemi del Mondo Tolemaico, e Copernicano* (Fiorenza: Gio. Batista Landini, 1632), 43 (trans. Martínez).

43. [Stobaeus], *Ioannis Stobaei Eclogarum Physicarum et Ethicarum Libri Duo,* ed., A. H. L. Heeren (Göttingen: Vandenhoek et Ruprecht, 1792), vol. 1, p. 17.

44. Burkert, *Lore and Science in Ancient Pythagoreanism,* 414–15.

45. See Diodorus, *The Historical Library of Diodorus the Sicilian, in Fifteen Books,* trans. G. Booth (London: J. Davis, 1814), vol. 1, book 1, p. 97.

46. Alan B. Loyd, introduction to *Herodotus Book 2* (Leiden: E. J. Brill, 1975), 47.

47. Mistranslation: that Moeris discovered the foundations of geometry and that Pythagoras "carried geometry to perfection"; see Diogenes Laertius, "Life of Pythagoras," sec. 11. For comments on these brief fragments ascribed to Hecataeus and Anticlides, see Felix Jacoby, ed., *Die Fragmente der grieschen Historiker* (Leiden: E. J. Brill, 1923–58), 264 f25.98.2, 140 f1, respectively.

48. Diodorus Siculus, *Bibliotheca Historica,* book 10, sec. 6 (trans. Martínez). One of the Oxyrhynchus Papyri (manuscripts from the first to the sixth century CE discovered in an ancient trash site, in Egypt) tells a story about Thales having been an "old man scraping the ground and drawing the figure discovered by the Phrygian Euphorbus, who was the first to draw even scalene triangles and a circle." So it would seem that Thales drew figures that had been discovered by an earlier incarnation of Pythagoras centuries before Pythagoras had been born: fiction. The translation here is from T. L Heath, *A Manual of Greek Mathematics* (Oxford: Oxford University Press, 1931), 92.

49. David C. Lindberg, *The Beginnings of Western Science,* 2nd ed. (Chicago: University of Chicago Press, 2007), 12.

50. Christiane L. Joost-Gaugier, *Measuring Heaven: Pythagoras and His Influence on Thought and Art in Antiquity and the Middle Ages* (Ithaca: Cornell University Press, 2006), 138, 248. Joost-Gaugier also claims that Aristotle praised Pythagoras as a mathematician interested in numbers (18), yet the purported evidence is only a phrase attributed to Apollonius (ca. 150 BCE). There is, however, "no reason whatever to think that the statement derives from Aristotle," as explained by W. A. Heidel in "The Pythagoreans and Greek Mathematics," *American Journal of Philology* 61, no. 1 (1940): 8.

51. Isidore Lévy, *Recherches sur les sources de la légende de Pythagore* (Paris: E. Leroux, 1926).

52. Among such texts are the "Golden Verses" (ca. 325 BCE?) and Lysis's letter to Hipparchus (ca. 225 BCE?); for discussion, see Riedweg, *Pythagoras,* 119–23.

53. Pliny the Elder, *Naturalis Historia* (ca. 79 CE), book 34, chap. 19.

54. Diogenes Laertius, "Life of Pythagoras," sec. 25.

55. Burkert, *Lore and Science in Ancient Pythagoreanism,* 415. See also Eva Sachs, *Die fünf Platonischen Körper* (Berlin: Weidmann, 1917).

56. O. Neugebauer, *The Exact Sciences in Antiquity,* in *Acta Historica Scientiarum Naturalium et Medicinalium,* ed. Jean Anker (Copenhagen: Ejnar Munksgaard, 1951), vol. 9, pp. 142, 143.

57. John Edgar Johnson, *The Monks before Christ: Their Spirit and Their History* (Boston: A. Williams and Company, 1870), 89.

58. Joy Hakim, *The Story of Science: Aristotle Leads the Way* (Washington, DC: Smithsonian Books, 2004), 81.

59. Eli Maor, *The Pythagorean Theorem: A 4,000-Year History* (Princeton: Princeton University Press, 2007), 17, 195; see also 24, 93, 186.

60. Stephen W. Hawking, *God Created the Integers* (Philadelphia: Running Press, 2005), 10; Thomas L. Heath, *The Thirteen Books of Euclid's Elements,* 3 vols., 2nd ed. (Cambridge: University Press, 1926), vol. 1, p. 353. This oversight in crediting Heath was corrected in the 2007 edition of Hawking's book.

61. Hardy Grant, "Back from Limbo: New Thoughts on the Mathematical

Philosophy of the Pythagoreans," *College Mathematics Journal* 34, no. 4 (2003): 343–48, quotation at 343.

2. An Irrational Murder at Sea

1. Plato, *Republic,* 531b–531c.

2. Aristotle, *Metaphysics,* book 1(A), chap. 5, lines 987b27–29.

3. Philolaus, quoted in Stobaeus (ca. 425 CE), *Eclogae,* 1.21.7b, known as Fragment 4, trans. and authenticated in Carl A. Huffman, *Philolaus of Croton, Pythagorean and Presocratic* (Cambridge: Cambridge University Press, 1993), 172–77.

4. For example, see Huffman, *Philolaus,* 147–48.

5. Aristotle, *Metaphysics,* book 1(A), chap. 5, line 985b.

6. Iamblichus, *On the Pythagorean Way,* chap. 12, p. 83.

7. Iamblichus, *On the Pythagorean Way,* chap. 28, p. 165.

8. Morris Kline, *Mathematical Thought from Ancient to Modern Times* (Oxford: Oxford University Press, 1972), vol. 1, p. 32. Likewise, in his bestselling book, Arthur Koestler wrote, "Hippasos, the disciple who let the scandal leak out, was put to death." See *The Sleepwalkers* (London: Hutchinson, 1959; reprint, London: Arkana/Penguin, 1989), 40.

9. William Everdell, *The First Moderns* (Chicago: University of Chicago Press, 1997), 33.

10. Simon Singh, *Fermat's Enigma: The Epic Quest to Solve the World's Greatest Mathematical Problem* (New York: Walker and Company, 1997), 50.

11. Charles Seife, *Zero: The Biography of a Dangerous Idea* (New York: Penguin, 2000), 26.

12. Douglas Hofstadter, *Gödel, Escher, Bach* (New York: Basic Books, 1979), 418, 556–57.

13. John Derbyshire, *Unknown Quantity: A Real and Imaginary History of Algebra* (Washington, DC: Joseph Henry Press, 2006), 199.

14. For example, Heath, *Thirteen Books of Euclid's Elements,* vol. 3, pp. 1–2. Yet Heath later admitted that "it is impossible that Pythagoras himself should have discovered a 'theory' or 'study' of irrationals in any proper sense." See Thomas Heath, *A History of Greek Mathematics* (Oxford: Clarendon Press, 1921), vol. 1, pp. 154–55.

15. I am astonished that some translators of Proclus's work freely choose to imagine that instead of the word ἀλόγων (irrational? unspeakable?), as Proclus's extant text seems to state, it originally read ἀναλόγων, so that it might now say that Pythagoras discovered "the doctrine of proportions." See, e.g., Proclus, *A Commentary on the First Book of Euclid's Elements,* trans. Glenn R. Morrow (Princeton: Princeton University Press, 1970), 53. Yet no documentary evidence indicates that Pythagoras contributed such a fundamental theory to mathematics either.

16. Kurt von Fritz, "The Discovery of Incommensurability by Hippasus of Metapontum," *Annals of Mathematics* 46, no. 2 (1945): 242–64.

17. Von Fritz did not specify an example, but see Paolo E. Arias, *Mille Anni di Ceramica Greca,* with photographs by M. Hirmer (Firenze: G. C. Sansoni, 1960), plate 14. The association of a star figure on a seventh-century BCE vase with the Pythagoreans is as arbitrary as claiming that they "may have" studied the swastika, since it too appeared on various vases; see, e.g., Arias, *Mille Anni,* plates 6, 10, 11, etc.

18. Lucian, *De lapsu in salutando* (ca. 180? CE), sec. 5, in *The Works of Lucian of Samosata,* trans. H. W. Fowler and F. G. Fowler (Oxford: Clarendon Press, 1905), vol. 2, "A Slip of the Tongue in Salutation," p. 36; and a later comment by an anonymous writer in a Scholium for an edition of the satirical play *Nubes,* by Aristophanes (ca. 423 BCE, which does not mention the Pythagoreans), line 609. Von Fritz does not give the date or edition for either of the works in question, but Thomas L. Heath, in *Thirteen Books of Euclid's Elements,* vol. 2, p. 99 (book 4, Proposition 10), refers to the source of the scholiast's comment: C. A. Bretschneider, *Die Geometrie und die Geometer vor Euklides, Ein Historischer Versuch* (Leipzig: B. G. Teubner, 1870), 85–86. In that work, however, Bretschneider does not give the date or the edition of Aristophanes's *Nubes,* although he notes that the scholiasts' comment is on line 611 (p. 249) and quotes the passage in Greek. The scholiast there claims (like Lucian) that by interlacing three triangles to make a pentagon, the Pythagoreans designated "health." I have not found a single writer who confronts the mathematical ambiguity of such words, namely that there are several distinct ways to construct either a pentagon or a pentagram using "three triangles."

19. For example, James R. Choike, "The Pentagram and the Discovery of an Irrational Number," *Two-Year College Mathematics Journal* 11, no. 5 (1980): 312–16.

20. Von Fritz, "Discovery of Incommensurability," 244–45; Zhmud, "Pythagoras as Mathematician," 257.

21. A. Wasserstein, "Theaetetus and the History of the Theory of Numbers," *Classical Quarterly, New Series* 8, nos. 3–4 (1958): 165–79, quotation at 165–66.

22. John Burnet, *Early Greek Philosophy* [1892], 4th ed. (London: Adam & Charles Black, 1930; reprint, 1948), 105–6.

23. Pappus of Alexandria, *Commentary on Book X of Euclid's Elements,* trans. from an Arabic text by G. Junge and W. Thomson (Cambridge, MA: Harvard University Press, 1930).

24. Iamblichus, *On the Pythagorean Way,* chap. 34, pp. 240–41; also in Iamblichus, *De Communi Mathematica,* 77.

25. Mario Livio, *The Golden Ratio* (New York: Broadway Books, 2006), 36.

26. Amir Alexander, *Duels at Dawn: Heroes, Martyrs and the Rise of Modern Mathematics* (Cambridge, MA: Harvard University Press, 2010), 7.

27. Frank Leslie Vatai, *Intellectuals in Politics in the Greek World* (London: Croom Helm, 1984), 147.

28. Iamblichus, *On the Pythagorean Way,* chap. 18, pp. 111–13. See also Iamblichus, *De Communi Mathematica,* 25, 77.

29. "Pythagoras, the son of Mnesarchus. . . . But some authors say that he was the son of Marmacus, the son of Hippasus, the son of Euthyphron, the son of Cleonymus" (Diogenes Laertius, "Life of Pythagoras," sec. 1).

30. Plato, *Theaetetus* (ca. 360 BCE), trans. Benjamin Jowett (New York: Charles Scribner's Sons, 1871), 10.

31. For discussion, see Wilbur R. Knorr, *The Evolution of the Euclidean Elements: A Study of the Theory of Incommensurable Magnitudes and Its Significance for Early Greek Geometry* (Boston: D. Reidel, 1975), 48.

32. Aristotle, *Prior Analytics* (ca. 350 BCE), i.23.41a26–27, trans. Ivor Thomas, in *Selections Illustrating the History of Greek Mathematics* (Cambridge, MA: Harvard University Press, 1939), 111.

33. Porphyry, *Life of Pythagoras* (trans. Guthrie), sec. 53, p. 133; Iamblichus, *On the Pythagorean Way,* chap. 31, p. 203.

34. Iamblichus, *On the Pythagorean Way,* chap. 17, pp. 99–101.

35. Hawking, *God Created the Integers,* 3.

36. Iamblichus, *On the Pythagorean Way,* chap. 32, p. 217.

37. Galileo Galilei, "Dialogo Primo," 3 (trans. Martínez).

3. Ugly Old Socrates on Eternal Truth

1. Antiphon to Socrates, according to Xenophon (a student of Socrates), in Xenophon, *Memorabilia* (composed ca. 371 BCE), book 1, chap. 6, in *The Works of Xenophon,* trans. H. G. Dakyns (London: Macmillan and Co., 1897), vol. 3, p. 32 (trans. modified).

2. Plato, *Apology of Socrates,* in *Socrates: Plato's Apology of Socrates and Crito, with a part of his Phaedo,* ed. and trans. Benjamin Jowett (New York: The Century Co., 1903), 63.

3. A. Plato, *Meno* (ca. 380 BCE), in Plato, *The Dialogues of Plato,* trans. Benjamin Jowett (Oxford: Clarendon Press, 1871).

4. Plato, *Republic* (ca. 375 BCE), trans. Benjamin Jowett (Oxford: Clarendon Press, 1888), book 7.

5. Aristotle, *Metaphysics,* book 1(A), chap. 6, line 987a 30.

6. Xenophon, *Memorabilia,* book 4, chap. 7, p. 176 (trans. modified).

7. Plato, *Apology of Socrates,* 72–73 (trans. modified).

8. Plato, *Apology of Socrates,* 79.

9. Plato, *Crito,* in Jewett, *Socrates,* 87–90, 114–15.

10. Plato, *Plato's Phædo,* trans. E. M. Cope (Cambridge: University Press, 1875), 78, 89–93, 102–8.

11. Joannes Philoponus, "Commentary on Aristotle's De Anima" (sixth c. CE),

in *In Aristotelis De Anima Libros Commentaria,* ed. Michael Hayduck, *Commentaria in Aristotelem Graeca* (Berlin: Reimer, 1897), vol. 15, pp. 117, 29; Elias, "Commentary on Aristotle's Analytics" (6th c. CE), in *In Categorias Commentarium,* ed. A. Busse, *Commentaria in Aristotelem Graeca,* part 1 (Berlin: Reimer, 1900), vol. 18, pp. 118, 18.

12. Plutarch, "Life of Marcellus" (ca. 100 A.D.), in *Plutarch's Lives of Illustrious Men,* trans. John Dryden and others, corrected from the Greek and rev. by Arthur H. Clough (1876; reprint, Boston: Little, Brown and Company, 1928), 221 (trans. modified).

13. Aristotle, quoted in Thomas L. Heath, ed., *Mathematics in Aristotle* (New York: Garland Publishing, 1980), 11.

14. Proclus, *Eis proton Eukleidou stoicheion biblon,* in Proclus, *Commentary.*

15. John Dee, "The Mathematicall Præface to Elements of Geometrie of Euclid of Megara," in *The Elements of Geometrie of the most auncient Philosopher Evclide of Megara,* trans. H. Billingsley (London: Iohn Daye, 1570), viij (spellings and punctuation modernized).

16. Dee, "Mathematicall Præface," viii.

17. Johannes Kepler, *Harmonices Mundi Libri V* (Lincii, Austria: Godofredi Tampachii, 1619), trans. by E. J. Aiton, A. M. Duncan, and J. V. Field as *The Harmony of the World* (Philadelphia: American Philosophical Society, 1997), 146.

18. Charles Hermite to Georg Cantor, quoted in Cantor to Hermite, 30 Nov, 1895, in Joseph Warren Dauben, *Georg Cantor: His Mathematics and Philosophy of the Infinite* (Princeton: Princeton University Press, 1990), 228; see also p. 348: "Hermite's original is unknown. It does not appear among the collection of papers and manuscripts of Cantor's surviving *Nachlass.*"

19. Kurt Gödel, "What Is Cantor's Continuum Problem?" in *Philosophy of Mathematics: Selected Readings,* ed. Paul Benacerraf and Hilary Putnam (1964), 2nd ed. (New York: Cambridge University Press, 1983), 258–73, reprinted in *Kurt Gödel Collected Works,* ed. Solomon Feferman, et al., vol. 2, *Publications 1938–1974* (New York: Oxford University Press, 1990), 254–70.

20. Kurt Gödel, "Some Basic Theorems on the Foundations of Mathematics and Their Implications," in Kurt Gödel, *Collected Works,* ed. Solomon Feferman, et al., vol. 3, *Unpublished Essays and Lectures* (Oxford: Oxford University Press, 1995), 304–23, quotations at 311–12.

21. I. R. Shafarevitch, "Über einige Tendenzen in der Entwicklung der Mathematik," Jahrbuch der Akademie der Wissenschaften in Göttingen (1973): German translation, 31–36; Russian original, 37–42. The English translation is from Philip J. Davis and Reuben Hersh, *The Mathematical Experience* (Boston: Birkhäuser, 1981), 52–54; see also Philip Merlan, *From Platonism to Neoplatonism* (The Hague: Martinus Nijhoff, 1960).

22. Philip J. Davis and Reuben Hersh, "Confessions of a Prep School Math Teacher," in *Mathematical Experience,* 272–74.

4. The Death of Archimedes

1. A work by Archimedes mentions Euclid, but historians agree that this lone mention is not original but was added by a later copyist. Still, Proclus mentioned that Archimedes cited Euclid.

2. Marcus Tullius Cicero, *De Oratore* (55 BCE), trans. by W. Guthrie as *The Three Dialogues of M. T. Cicero* (New York: Harper & Brothers, 1857), book 3, sec. 33, p. 309: "when Euclid or Archimedes taught geometry."

3. Pappus, *Synagoge* (ca. 320 CE), in Pappus, *Collection: Book 7,* ed. Alexander Jones (New York: Springer, 1986).

4. Proclus, *Commentary,* 57.

5. Proclus, *Commentary,* 57.

6. Heath, *Thirteen Books of Euclid's Elements,* 1: 225. Heath referred specifically to propositions 1.4 and 1.8.

7. Heath did not cite references, but see Bertrand Russell, *An Essay on the Foundations of Geometry* (Cambridge: University Press, 1897), 67–88; Arthur Schopenhauer, *Die Welt als Wille und Vorstellung,* vol. 1 (1844), trans. by R. Haldane and J. Kemp as *The World as Will and Idea,* vol. 1 (London: Trübner & Co. 1883), 36, 70, 159.

8. Proclus, *A Commentary on the First Book of Euclid's Elements,* ed. Glenn Raymond Morrow (Princeton: Princeton University Press, 1992), 46.

9. Archimedes, "On Spirals," in *The Works of Archimedes,* ed. Thomas L. Heath (Cambridge: University Press, 1897), 154.

10. Plutarch, "Life of Marcellus," 223 (trans. modified).

11. Archimedes, "On the Equilibrium of Planes," in *Works,* 189.

12. Ernst Mach, *Die Mechanik in ihrer Entwickelung Historisch-Kritisch Dargestellt* (Leipzig: F. A. Brockhaus, 1883), 2nd. ed. trans. by Thomas J. McCormack (1889) as *The Science of Mechanics: A Critical and Historical Account of Its Development* (Chicago: Open Court, 1893), 9.

13. Archimedes, *The Method of Archimedes* [ca. 225 BCE] *Recently Discovered by Heiberg; a supplement to the Works of Archimedes,* trans. Thomas L. Heath (Cambridge: Cambridge University Press, 1912).

14. Reviel Netz, Ken Saito, and Natalie Tchernetska, "A New Reading of Method Proposition 14: Preliminary Evidence from the Archimedes Palimpsest: Part 1," *Sciamus* 2 (2001): 9–29, and "Part 2," *Sciamus* 3 (2002): 109–25.

15. Reviel Netz and William Noel, *The Archimedex Codex* (Philadelphia: Da Capo Press, 2007), 201. Reviel Netz, William Noel, Natalie Tchernetska, and Nigel Wilson, eds., *The Archimedes Palimpsest,* 2 vols. (Cambridge: Cambridge University Press, 2011).

16. Reviel Netz, Fabio Acerbi, and Nigel Wilson, "Towards a Reconstruction of Archimedes' Stomachion," *Sciamus* 5 (2004): 67–99.

17. Vitruvius, *De Architectura* (ca. 15 BCE), trans. by Morris Hicky Morgan as *The*

Ten Books of Architecture (Cambridge, MA: Harvard University Press, 1914), book 9, pp. 253–54 (trans. modified).

18. Galileo Galilei, "La Bilancetta" (1586), in *Opera di Galileo Galilei,* ed. Franz Brunetti (Torino: Unione Tipografico-Editrice Torinese, 1980) (trans. Martínez).

19. Plutarch, "Life of Marcellus," 221 (trans. modified).

20. Plutarch, "Life of Marcellus," 222 (trans. modified).

21. Plutarch, "Life of Marcellus," 223 (trans. Martínez).

22. Marcus Tullius Cicero, *Against Verres,* book 2, sec. 4.131 (trans. Martínez), also in *Cicero: Select Orations* (New York: Harper & Brothers, 1889), 571; Marcus Tullius Cicero, *De Finibus,* book 5, sec. 50 (trans. Martínez), also in *On Moral Ends,* ed. Julia Annas (Cambridge: Cambridge University Press, 2001); Livy, *History of Rome from Its Foundation,* book 25, sec. 31 (trans. Martínez), also in Aubrey de Selincourt, *The War with Hannibal* (New York: Penguin, 1965), 338; Valerius Maximus, *Memorable Doings and Sayings,* book 8, sec. 7.7, Loeb Classical Library (Cambridge, MA: Harvard University Press, 2000), vol. 2, books 6–9 (trans. Martínez); Plutarch, "Life of Marcellus," 221 (trans. Martínez); John Tzetzes (ca. twelfth c. AD), *Book of Histories* (Chiliades), book 2, lines 136–49 (trans. Martínez), also in Ivor Thomas, *Greek Mathematical Works,* Loeb Classical Library (Cambridge, MA: Harvard University Press, 1941), vol. 2; John Zonaras (ca. twelfth c. AD), *Epitome ton Istorion,* 9, 5 (trans. Martínez), also in Earnest Cary, *Dio's Roman History,* vol. 2, *Fragments of Books XII–XXV,* Loeb Classical Library (Cambridge, MA: Harvard University Press, 1914). See also, http://www .math.nyu.edu/~%20crorres/Archimedes/Death/Histories.html.

23. Bartholomeo Keckermanno, *Systema Compendiosvm Totivs Mathematices, hoc est, Geometriæ, Opticæ, Astronomiæ et Geographiæ* (Hanoviæ: Petrvm Antonivs, 1621), Prolegomena Scientiavm Mathematicarvm, 37. See also Andreas Lazarus von Imhof, *Le Grand Théâtre Historique, ou Nouvelle Histoire Universelle, tant Sacrée que Profane . . .* (Leide: Pierre Vander, 1703), vol. 1, chap. 8, p. 404: "Miles, noli turbare meos circulos"—i.e., "Soldier, do not disturb my circles."

24. Bell, *Men of Mathematics,* 34; Petr Beckmann, *A History of π (PI),* 3rd ed. (New York: Barnes & Noble, 1993), 61; Kline, *Mathematical Thought,* vol. 1, p. 106; Hawking, *God Created the Integers,* 121; Paul Hoffman, *Archimedes' Revenge* (New York: W. W. Norton, 1988), 27; Timothy Ferris, *Coming of Age in the Milky Way,* new ed. (New York, HarperCollins, 2003), 40; Maor, *Pythagorean Theorem,* 50–51; Alexander, *Duels at Dawn,* 7.

25. For example: Alan Hirshfeld, *Eureka Man: The Life and Legacy of Archimedes* (New York: Walker Publishing Co., 2009), 153; Serafina Cuomo, *Ancient Mathematics* (London: Routledge, 2001), 200; William Dunham, *Journey through Genius: The Great Theorems of Mathematics* (New York: Penguin Books, 1991), 88.

26. Marcus Tullius Cicero, *Tusculanae Disputationes* (ca. 45 BCE), in *Tusculan Disputations; also, Treatises on the Nature of Gods, and on the Commonwealth,* trans.

C. D. Yonge (New York: Harper & Brothers, 1899), 186–87. He continued: "For you must necessarily look there for the best of every thing, where the excellency of man is; but what is there better in man than a sagacious and good mind?"

5. Gauss, Galois, and the Golden Ratio

1. Justinus Martyr [apocryphal?], *Cohort ad Græcos* (ca. 150 CE), chap. 31, in *The Writings of Justin Martyr and Athenagoras,* trans. Marcus Dods, George Reith, and B. Pratten, vol. 2 of *Ante-Nicene Christian Library: Translations of the Writings of the Fathers down to A.D. 325,* ed. Alexander Roberts and James Donaldson (Edinburgh: T. &. T. Clark, 1868), 415.

2. Margaret B. W. Tent, *The Prince of Mathematics: Carl Friedrich Gauss* (Wellesley, MA: A. K. Peters, 2006), 33–34; Karin Reich, *Carl Friedrich Gauss: 1777-1977* (Bad Godesberg: Inter Nationes, 1977), 7–8; Tord Hall, *Carl Friedrich Gauss: A Biography* (Cambridge, MA: MIT Press, 1970), 3-5; Guy Waldo Dunnington, *Carl Friedrich Gauss, Titan of Science: A Study of His Life and Work* (New York: Exposition Press, 1955), 12.

3. Tony Rothman, *Everything's Relative: And Other Fables from Science and Technology* (New Jersey: John Wiley & Sons, 2003), 233.

4. Brian Hayes, "Gauss's Day of Reckoning: A Famous Story about the Boy Wonder of Mathematics Has Taken on a Life of its Own," *American Scientist, The Magazine of Sigma Xi, the Scientific Research Society* 94 (May–June 2006): 200–205.

5. W. Sartorius von Walthershausen, *Gauss: A Memorial* (1856), trans. Helen Worthington Gauss (Colorado Springs: s.n., 1966?); this translation has defects, as noted in Hayes, "Gauss's Day of Reckoning," 201.

6. Bell, *Men of Mathematics,* 221.

7. For example, Steven George Krantz, *Mathematical Apocrypha Redux: More Stories and Anecdotes of Mathematicians* (Cambridge: Cambridge University Press, 2005), 231.

8. For example, Steve Olson, *Count Down: Six Kids Vie for Glory at the World's Toughest Math Competition* (New York: Mariner/Houghton Mifflin, 2004), 67; Marcus Du Sautoy, *The Music of the Primes: Searching to Solve the Greatest Mystery in Mathematics* (New York: HarperCollins, 2004), 25; John Derbyshire, *Prime Obsession: Bernhard Riemann and the Greatest Unsolved Problem in Mathematics* (Washington, DC: Joseph Henry Press, 2003), 48; Theoni Pappas, *The Joy of Mathematics,* rev. ed. (San Carlos, CA: Wide World Publishing/Tetra, 1989), 164. The earliest instance of the sum of 1 to 100, found by Brian Hayes, is Franz Mathé, *Karl Friedrich Gauss* (Leipzig: Im Feuer Verlag, 1906), 3–4.

9. Charles Stanley Ogilvy and John T. Anderson, *Excursions in Number Theory* (New York: Oxford University Press, 1966; New York: Dover, 1988), 12.

10. Martin Goldstein and Inge F. Goldstein, *The Experience of Science: An Interdisciplinary Approach* (New York: Plenum Press, 1984), 89.

11. Hayes, "Gauss's Day of Reckoning," 203.

12. Bell, *Men of Mathematics,* 362.

13. Tony Rothman, "Genius and Biographers: The Fictionalization of Evariste Galois," *American Mathematical Monthly* 89, no. 2 (1982): 84–106.

14. Leopold Infeld, *Whom the Gods Love: The Story of Evariste Galois* (New York: Whittlesey House, 1948), 308–11.

15. Fred Hoyle, *Ten Faces of the Universe* (San Francisco: W. H. Freeman, 1977), 14–15.

16. Tony Rothman, "Genius and Biographers: The Fictionalization of Evariste Galois," expanded version, http://www.physics.princeton.edu/~trothman/galois.html. For an account based on primary sources, see Laura Toti Rigatelly, *Evariste Galois, 1811–1832* (Basel: Birkhäuser, 1993), 13–14, 104–14. Toti Rigatelli argues that Galois planned to die in what was actually a false duel so that his death could serve republicans as a means to ignite a public uprising against King Louis-Philippe, but then the subsequent death of a general gave the republicans a more popular icon to rally around.

17. E. T. Bell to John Macrae, 1928, quoted in Reid, *Search for E. T. Bell,* 240.

18. Reid, *Search for E. T. Bell,* 290.

19. Sophie Germain to Guglielmo Libri, 18 Apr. 1831, in C. Henry, "Les manuscrits de Sophie Germain et leur récent éditeur," *Revue Philosophique de la France et de l'Étranger* 8 (July–Dec. 1831): 632 (trans. Martínez).

20. Paul Dupuy, "La vie d'Évariste Galois," *Annales Scientifiques de l'Ecole Normale Supérieure* 13 (1896): 197–266, esp. 234–35.

21. Francois Vincent Raspail, *Lettres sur les prisons de Paris* (Paris: 1839), vol. 2, p. 90.

22. Évariste Galois, preface to *Deux mémoires d'analyse pure* (1831), reproduced in René Taton, "Les relations d'Évariste Galois avec les mathématiciens de son temps," *Revue d'Histoire des Sciences et de leurs Applications* 1, nos. 1–2 (1947): 114–30, quotations at 123–25 (trans. Martínez).

23. Vitruvius, *De Architectura* (Morgan), introduction to book 7, p. 197.

24. Miguel de Cervantes Saavedra, prologue to *El Ingenioso Hidalgo Don Quixote de la Mancha* (Madrid: Juan de la Cuesta, 1605), 2; *maldiciente* means badmouthed, a slanderer or damner.

25. James Wood, *The Nuttal Encyclopædia, being a Concise and Comprehensive Dictionary of General Knowledge* (London: Frederick Warne and Co., 1920), 699.

26. Galois, preface, 123 (trans. Martínez).

27. Galois, preface, 123–25 (trans. Martínez).

28. Évariste Galois, manuscript, "14 mai 83," in *Manuscrits de Évariste Galois,* ed.

Jules Tannery (Paris: Gauthier-Villars, 1908), 3 (ellipses corresponding to blank spaces in Tannery's edition) (trans. Martínez).

29. Carlos Alberto Infantozzi, "Sur la mort d'Évariste Galois," *Revue d'Histoire des Sciences et de leurs Applications* 21, no. 2 (1968): 157–60; the manuscripts of Galois were located in Paris at the Bibliothèque de l'Institut.

30. Infantozzi, "Sur la mort d'Évariste Galois," 159.

31. Évariste Galois, manuscript, n.d., in Galois, *Manuscrits,* 3–4 (ellipses corresponding to blank spaces in the originals and in Tannery's edition) (trans. Martínez). Cf. Rothmans's translation of this letter; e.g., he includes the phrase "but it has been very wounded."

32. Évariste Galois to Auguste Chevalier, 25 May 1832, in *Ecrits et memoires mathematiques d'Évariste Galois: Edition critique integrale de ses manuscrits et publica-tions,* ed. Robert Bourgne and J. P. Azra (Paris: Gauthier- Villars, 1962), 468–69.

33. Évariste Galois to his republican friends, 29 May 1832, *Revue Encyclopédique* 55 (Sept. 1832): 753 (trans. Martínez).

34. Évariste Galois to two friends, 29 May 1832, in *Revue Encyclopédique* 55 (Sept. 1832): 753–54 (trans. Martínez).

35. Évariste Galois to Auguste Chevalier, 25 May 1832, in *Revue Encyclopédique* 55 (July 1832): 576 (trans. Martínez).

36. *Le Précurseur* (Lyon), 4 June 1832 (trans. Martínez).

37. Rothman, "Genius and Biographers" (expanded version), sec. 5.

38. Luca Pacioli, *Divina Proportione: Opera a Tutti Glingegni Perspicaci e Curiosi Necessaria que Ciascun Studioso di Philosophia: Prospectiva Pictura, Sculptura, Architec-tura, Musica, e altre Mathematice: Suavissima, Sottile, e Admirabile Doctrina consequira, e Delectarassi: con varie Questione de Secretissima Scientia* (Venetiis: A. Paganius Paganinus, 1509), chap. 5 (trans. Martínez).

39. Kepler, *Harmony of the World,* 31.

40. Ioanne Keplero, *Prodromus Dissertationvm Cosmographicarvm, continens Mysterivm Cosmographicvm de Admirabili Proportione Orbium Cœlestium,* 2nd ed. (Francofvrti: Erasmi Kempferi/Godefridi Tampachii, 1621), chap. 12, pp. 42, 47 (trans. Martínez); these passages do not appear in the original edition of 1596.

41. P. Ignatio Pickel, *Elementa Arithmeticae, Algebrae, ac Geometriae* (Dilingæ: Joannis Leonard, 1772), vol. 2, p. 39.

42. Aegidius Mosmayr, *Geistliches und allgemeines Befreyungs-Ort* (Mergentheim: Quirini Hehl, 1695), 74; Dr. Johann Theodor Ellers, *Vollständige Chirurgie: oder, Gründliche Anweisung alle und jede ausserliche Krankheiten des menschlichen Körpers* (Berlin: Gottlieb August Lange, 1763), 698.

43. I. Ritter von Seyfried, ed., *Allgemeine Musikalisch Zeitung, mit besonderer Rücksicht auf den österreichischen Kaiserstaat* 4, no. 70 (30 Aug. 1820): 555: "der bekannten *Sectio divina geometrica* (auch schlechthin *Sectio divina* oder *Sectio aurea . . .*)."

44. Ephraim Salomon Unger, *Praktische Uebungen für angehende Mathematiker. Ein Hülfsbuch für Alle, welche die Fertigkeit zu erlangen wünschen, die Mathematik mit Nutzen anwenden zu können* (Leipzig: F. U. Brockhaus, 1828), vol. 1, p. 204: "Da nun dieses die unter dem Namen der goldene Schnitt bekannte Aufgabe ist, welche im 11ten Satze des zweiten Buchs der Elemente von Euklid gelöst wird." See also Ferdinand Wolff, *Lehrbuch der Geometrie. Erster Theil, Elemenar: Geometrie und ebene Trigonometrie* (Berlin: A. Petsch 1830), 160: "goldenen Schnitt."

45. Adolf Zeising, *Neue Lehre von den Proportionen des Menschlichen Körpers, aus einem bisher unerkannt gebliebenen, die ganze Natur und Kunst durchdringenden morphologischen Grundgesetze entwickelt und mit einer vollständigen historischen Uebersicht der bisherigen Systeme* (Leipzig: Rudolph Weigel, 1854), 10; see also 14, 24.

46. Zeising, *Neue Lehre*, 178–81 (trans. Martínez).

47. Book review: "Neue Lehre von den Proportionen des Menschlichen Körpers," *British Quarterly Review* 21 (Jan. 1855): 267–68.

48. Pacioli, *Divina Proportione*, chap. 18, folio 32 verso; more specifically, Pacioli recommended the use of geometric figures, which "not only will serve as decoration but also will give scholars and wise men some motives to speculate, being always constructed with that sacred and divine proportion" (trans. Martínez).

49. Zeising, *Neue Lehre*, 393-409; he specifies churches in Germany: the Cologne Cathedral and St. Elizabeth in Marburg/Hesse.

50. Gustav Theodor Fechner, "Ueber die Frage des goldenen Schnittes," *Archiv für die zeichnenden Künste* 11 (1865): 100–112.

51. Gustav Theodor Fechner, *Vorschule der Ästhetik* (Leipzig: Breitkopf & Härtel, 1876), 190–99.

52. Fechner, "Ueber die Frage des goldenen Schnittes," 162; Gustav Theodor Fechner, "Warum wird die Wurst schief durchgeschnitten?" in *Kleine Schriften* (Leipzig: Breitkopf & Härtel, 1913), 255–70.

53. Theodore A. Cook, "A New Disease in Architecture," *Nineteenth Century and After* 91 (1922): 521–53, quotation at 521.

54. For example, see David M. Burton, *The History of Mathematics: An Introduction* (Boston: Allyn and Bacon, 1985), 62; Herbert Western Turnbull, "The Great Mathematicians," in *The New World of Mathematics,* ed. James R. Newman (New York: Simon and Schuster, 1956), 80.

55. As rightly shown in Gillings, *Mathematics in the Time of the Pharaohs*, 238–39, and in Roger Fischler, "What Did Herodotus Really Say? or, How to Build (a Theory of) the Great Pyramid," *Environment and Planning B* 6 (1979): 89–93.

56. George Markowsky, "Misconceptions about the Golden Ratio," *College Mathematics Journal* 23, no. 1 (1992): 2–19, esp. 6–8.

57. R. Fischler, "Théories mathématiques de la Grande Pyramide," *Crux Mathematicorum* 4 (1978): 122–29. For an excellent study of the actual dimensions of the Great Pyramid, which discusses various myths and theories, see Roger Herz-

Fischler, *The Shape of the Great Pyramid* (Waterloo, ON: Wilfred Laurier University Press, 2000). (Roger Fischler began to publish as Herz-Fischler in 1982.)

58. Marcel Dansei, *The Puzzle Instinct: The Meaning of Puzzles in Human Life* (Bloomington: Indiana University Press, 2002), 101.

59. *Donald in Mathmagic Land,* by Milt Banta, Bill Berg, Heinz Haber, narr. Paul Frees, prod. Walt Disney (Walt Disney Productions, 1959).

60. Three more examples among many: H. E. Huntley, *The Divine Proportion* (New York: Dover, 1970), 25, 62–63; Pappas, *Joy of Mathematics,* 102; George Manuel and Amalia Santiago, "An Unexpected Appearance of the Golden Ratio," *College Mathematics Journal* 19 (1988): 168–70, esp. 168.

61. Markowsky, "Misconceptions."

62. David Bergamini and the editors of LIFE, *Mathematics* (New York: Time, 1963), 94, 96.

63. Markowsky, "Misconceptions," 10–12; Roger Fischler, "The Early Relationship of Le Corbusier to 'the Golden Number,'" *Environment and Planning B* 6 (1979): 95–103; Roger Fischler, "On the Application of the Golden Ratio to the Visual Arts," *Leonardo* 14 (1981): 31–32.

64. Margaret F. Willerding, *Mathematical Concepts: A Historical Approach* (Boston: Prindle, Weber & Schmidt, 1980), 74; Frank Land, *The Language of Mathematics* (Garden City, NY: Doubleday, 1963), 222.

65. For example, see H. Schiffman and D. Bobko, "Preference in Linear Partitioning: the Golden Section Reexamined," *Perception and Psychophysics* 24 (1978): 102–3. See also Markowsky, "Misconceptions," 13–15.

66. The Internet poll was at http://homepage.esoterica.pt/~madureir (no longer available).

67. Malcolm W. Browne, "'Impossible' Form of Matter Takes Spotlight in Study of Solids," *New York Times,* 5 Sept. 1989, C1, C11.

68. Lori K. Calise, Tamara M. Caruso, John B. Cunningham, and Paul M. Sommers, "The Golden Midd," *Journal of Recreational Mathematics* 24, no. 1 (1992): 26–29.

69. Martin Gardner, "The Cult of the Golden Ratio," *Skeptical Inquirer* 18, no. 3 (1994): 243–45.

70. John F. Putz, "The Golden Section and the Piano Sonatas of Mozart," *Mathematics Magazine* 68, no. 4 (1995): 275–81.

71. Three examples among many: Theoni Pappas, *More Joy of Mathematics* (San Carlos, CA: World Wide Publishing/Tetra, 1991), 56–57; Seife, *Zero,* 32; Livio, *Golden Ratio,* 9. Livio writes: "The growth of spiral shells also obeys a pattern that is governed by the Golden Ratio." The covers of the books by Pappas, Livio, and others are decorated with illustrations of a chambered nautilus shell.

72. Clement Falbo, "The Golden Ratio: A Contrary Viewpoint," *College Mathematics Journal* 36, no. 2 (2005): 123–34, esp. 127. The pedagogical challenge

remains: how should we replace the traditional false stories about the golden ratio with something that is better and true? I suggest that we can well say: like other irrational numbers, the golden ratio has *never* been found to be embodied in the length of any natural or artificial object. Like Euclid's triangles, it seems to exist only in the imagination, yet it too is immensely useful in practical realms, because its value is roughly close to certain proportions among physical quantities.

73. Dan Brown, *The Da Vinci Code* (Doubleday: New York, 2003), 93–96.

74. D. Brown, *Da Vinci Code*, 95.

75. Dan Brown, in "Interview with Dan Brown," interviewed by Martin Savidge, *CNN Sunday Morning*, 25 May 2003, http://transcripts.cnn.com/TRANSCRIPTS/0305/25/sm.21.html.

76. Ethan Watrall and Jeff Siarto, *Head First Web Design* (Sebastopol, CA: O'Reilly Media, 2009), 133; Ivan Moscovitch, *The Hinged Square and Other Puzzles* (New York: Sterling Publishing Co., 2004), 76, 122; Ralph A. Clevenger, *Photographing Nature: A Photo Workshop from Brooks Institute's Top Nature Photography Instructor* (Berkeley: New Riders, 2010), 238; Doug Sahlin, *Digital Portrait Photography for Dummies* (Hoboken, NJ: Wiley, 2010), 99; Alexander R. Cuthbert, *The Form of Cities: Political Economy and Urban Design* (Malden, MA: Blackwell Publishing, 2006), 175; George Alexander MacLean, *Fibonacci and Gann Applications in Financial Markets* (West Sussex: John Wiley & Sons, 2005), 5; Tom Bierovic, *Playing for Keeps in Stocks and Futures* (New York: John Wiley & Sons, 2002), 63; Kurt Brown, *The Measured Word: On Poetry and Science* (Athens: University of Georgia Press, 2001), 167; Sarah Susanka, *Creating the Not So Big House* (Newtown, CT: Taunton Press, 2001), 103.

77. Hans Walser, *Der Goldene Schnitt,* 2nd ed. (Stuttgart: Teubner, 1996), trans. by Peter Hilton, with Jean Pedersen, as *The Golden Section* (Washington, DC: Mathematical Association of America, 2001).

78. Livio, *Golden Ratio,* 35.

79. Scott Olsen, *The Golden Section: Nature's Greatest Secret* (New York: Walker Publishing Company, 2006).

80. Livio, *Golden Ratio,* 6.

81. Seife, *Zero,* 32.

6. From Nothing to Infinity

1. Aristotle, *Physics,* book 4, chap. 8, line 215b, in *The Works of Aristotle,* ed. W. D. Ross and J. A. Smith (Oxford: Clarendon Press, 1908), vol. 2; Carl B. Boyer, "An Early Reference to Division by Zero," *American Mathematical Monthly 50*, no. 8 (1943): 487–91.

2. Henry T. Colebrooke, *Algebra with Arithmetic and Mensuration from the Sanscrit of Brahmagupta and Bháscara* (London, 1817), sec. 35–36, pp. 339–40.

3. Johannis Wallisii, *Arithmetica Infinitorum: sive Nova Methodus inquirendi in*

Curvilineorum Quadraturam, aliaq, difficiliora Matheseos Problemata (Oxford: Leon. Lichfield, 1656), 74–75, 87; Johannis Wallisii, *De Sectionibus Conicis, Nova Methodo Expositis, Tractatus* (Oxonii: Tho. Robinson, 1655), 8.

4. Leonhard Euler, *Vollständige Anleitung zur Algebra,* part 1 (St. Petersburg: Kays. Acad. der Wissenschaften, 1770, 1771), 34; Euler, *Élémens d'algebre,* trans. J. Bernoulli, with additions by J.-L. Lagrange (Lyon: Bryset & Co., 1784), 59–60; Euler, *Elements of Algebra,* ed. J. Hewlett, trans. from the French by Francis Horner, 2nd ed. (London: J. Johnson and Co., 1810), 34.

5. Martin Ohm, *Arithmetik und Algebra*, vol. 1 of *Versuch eines Vollkommen Consequenten Systems der Mathematik* (Berlin: I. H. Riemann, 1828); chap. 3, sec. 95, p. 112 (trans. Martínez).

6. Martin Ohm, *Der Geist der Mathematischen Analysis und ihr Verhältniss zur Schule* (Berlin: Dunder und Humblot, 1842), vol. 1, p. 40; Martin Ohm, *The Spirit of Mathematical Analysis, and Its Relation to a Logical System,* trans. Alexander John Ellis (London: John W. Parker, 1843), 26.

7. See, e.g., William Anthony Granville, *Elements of the Differential and Integral Calculus* (Boston: Ginn & Company, 1904), 10.

8. Constance Reid, *From Zero to Infinity,* 4th ed. (Washington, DC: Mathematical Association of America, 1992), 15.

9. Sylvestre François Lacroix, *Élémens d'algèbre: a l'usage de l'École Centrale des Quatre-Nations* (Berlin: I. H. Riemann, 1828); chap. 3, sec. 95, p. 112; Ohm, *Arithmetik und Algebra,* 112; George Peacock, *A Treatise on Algebra* [vol. 1] (Cambridge: J. & J. J. Deighton, 1830), 281–82; Augustus De Morgan, *On the Study and Difficulties of Mathematics,* Library of Useful Knowledge, vol. 1 (London, 1831), 41–42; Charles Smith, *Elementary Algebra for the Use of Preparatory Schools,* rev. Irving Stringham (London: Macmillan Co., 1896), 204, 252; Louis Couturat, *De l'infini mathématique* (Paris: Germer Baillière/Félix Alcan, 1896), 93–95; Edwin Bidwell Wilson, *Advanced Calculus: A Text upon Select Parts of Differential Calculus, Differential Equations, Integral Calculus, Theory of Functions, with Numerous Exercises* (Boston: Ginn and Company, 1911), 41, 258; Konrad Knopp, *Theory and Application of Infinite Series,* vol. 2, trans. R. C. Young (London: Blackie & Son, 1928), 5.

10. Leonhard Euler, "Recherches sur les racines imaginaires des equations (1749)," *Memoires de l'Académie des Sciences de Berlin* 5 (1751), reprinted in Leonhardi Euleri, *Opera Omnia,* First Series: *Opera Mathematica* (Lipsiae et Berolini: B. G. Teubneri, 1921), vol. 6, pp. 80–81, 113 (page numbers refer to *Opera* reprint); Euler, *Vollständige Anleitung zur Algebra* (1770), 88.

11. Hieronymi Cardani, *Ars Magna; Artis Magnae, sive, De Regvlis Algebraicis, Liber Unus: qui & Totius Operis de Arithmetica* (Norimbergae: Ioh. Petreium, 1545), ed. and trans. by T. Richard Witmer as *Girolamo Cardano, The Great Art, or the Rules of Algebra* (Cambridge, MA: MIT Press, 1968).

12. Jean d'Alembert, "Sur le les quantités négatives," *Opuscules Mathématiques 8* (1780): 277 (trans. Martínez).

13. Morris Kline, "Arithmetics and Their Algebras," in *Mathematics for Liberal Arts* (Reading, MA: Addison-Wesley, 1967), reissued as *Mathematics for the Non-Mathematician* (New York: Dover Publications, 1985), 478–98.

14. James Anderson, speaking in a video featured in Ben Moore (reporter), "1200-Year-Old Problem 'Easy': Schoolchildren from Caversham Have Become the First to Learn a Brand New Theory That Dividing by Zero Is Possible Using a New Number—'Nullity.' But the Suggestion Has Left Many Mathematicians Cold," *BBC South Today,* 12 Dec. 2006, http://www.bbc.co.uk/ berkshire/content/ articles/2006/12/06/divide_zero_feature.shtml.

15. In the video, Dr. Anderson teaches the subject to students at Highdown School, in Emmer Green. See Moore, "1200-Year-Old Problem 'Easy.'" The article ends with the lines: "Despite being a problem tackled by the famous mathematicians Newton and Pythagoras without success, it seems the Year 10 children at Highdown now know their nullity."

16. Amir D. Aczel, *The Mystery of the Aleph: Mathematics, the Kabbalah, and the Search for Infinity* (New York: Four Walls Eight Windows, 2000), 13.

17. "Sunk by Windows NT," *Wired Magazine,* online ed., 24 July 1998, http:// www.wired.com/science/discoveries/news/1998/07/13987.

18. Richard Courant and Herbert Robbins, *What Is Mathematics? An Elementary Approach to Ideas and Methods* (New York: Oxford University Press, 1941), 2nd ed., rev. Ian Stewart (New York: Oxford University Press, 1996), 15 (notation for multiplication modified slightly, for clarity).

19. John Stillwell, *The Four Pillars of Geometry* (New York: Springer, 2005), 106. See also Andreas Enge, *Elliptic Curves and Their Applications to Cryptography* (New York: Springer, 1999), 27; John Stillwell, *Mathematics and Its History,* 3rd ed. (New York: Springer, 2010), 146. See also the following: K. G. Binmore, *The Foundations of Analysis: A Straightforward Introduction. Book 2: Topological Ideas* (Cambridge: Cambridge University Press, 1981), 165; Stephanie Frank Singer, *Linearity, Symmetry, and Prediction in the Hydrogen Atom* (New York: Springer, 2005), 301; David E. Alexander, *Nature's Flyers: Birds, Insects, and the Biomechanics of Flight* (Baltimore: Johns Hopkins University Press, 2002), 152.

20. Om P. Chug, P. N. Gupta, and R. S. Dahiya, *Topics in Mathematics Calculus and Ordinary Differential Equations* (New Delhi: Laxmi Publications, 2008), 4.

21. Bernard W. Taylor, *Introduction to Management Science,* 9th ed. (New Jersey: Pearson Education/Prentice Hall, 2007), "Module A: The Simplex Solution Method," 26–27.

22. For example, see William D. Stanley, *Technical Analysis and Applications with MATLAB* (New York: Thomson Learning, 2007), 5.

23. James Leiterman, *Learn Vertex and Pixel Shader Programming with DirectX 9* (Plano, TX: Wordware Publishing, 2004), 68–69.

24. References to books are given in other footnotes. See also Arnold Emch, "On Limits," *American Mathematical Monthly* 9, no. 1 (1902): 5–9, esp. 6.

25. Robert Donald Smith, *Mathematics for Machine Technology,* 5th ed. (New York: Thomson Learning, 2004), 383; Robert D. Smith and John C. Peterson, *Introductory Technical Mathematics* (New York: Thomson Learning, 200), 700.

26. Eli Maor, "Thou Shalt Not Divide by Zero!" *Math Horizons* 11, no. 2 (2003): 16–19.

27. Alberto A. Martínez, *Negative Math: How Mathematical Rules Can Be Positively Bent* (Princeton: Princeton University Press, 2005), 18–67.

28. For example, if the binomial theorem is to hold for at least one nonnegative integer exponent, then zero to the power of zero should be 1. For some history see Donald E. Knuth, "Two Notes on Notation," *American Mathematical Monthly* 99, no. 5 (1992): 403–22.

29. IEEE data from 31 Dec. 2010 at http://www.ieee.org/about/today/at_a_glance.html#sect1.

30. James A. D. W. Anderson, Norbert Völker, and Andrew A. Adams, "Perspex Machine VIII: Axioms of Transreal Arithmetic," in *Proceedings of the IS&T/SPIE Conference on Vision Geometry XV,* vol. 6499, ed. Longin Jan Latecki, David M. Mount, and Angela Y. Wu (San Jose: SPIE, 2007), 649902.

31. James A. D. W. Anderson, "Introduction," http://www.bookofparagon.com.

32. Charles Kittel and Herbert Kroemer, *Thermal Physics,* 2nd ed. (W. H. Freeman Company, 1980), 462.

33. Seife, *Zero,* 214.

7. Euler's Imaginary Mistakes

1. Martínez, *Negative Math.*

2. Morris Kline, *Mathematics: The Loss of Certainty* (New York: Oxford University Press, 1980), 121; Paul Nahin, *An Imaginary Tale: The Story of √-1* (Princeton: Princeton University Press, 1998), 12.

3. For a history of some of the controversies surrounding "impossible" numbers before the 1750s, see Helena Pycior, *Symbols, Impossible Numbers, and Geometric Entanglements* (Cambridge: Cambridge University Press, 1997).

4. François Daviet de Foncenex, "Reflexions sur les quantités imaginaires," *Miscellanea Philosophico-Mathematica Societatis Taurinensis* 1 (1765): 113–46.

5. Jean d'Alembert, "Sur les logarithmes des quantités négatives, et supplément" (1759), in *Opuscules mathématiques* (Paris: David, 1761), vol. 1, pp. 180–230.

6. For a discussion of how, in connection with Euler's work, various kinds of numbers were considered fictitious instead of real quantities, see G. Ferraro,

"Differentials and Differential Coefficients in the Eulerian Foundations of the Calculus," *Historia Mathematica* 31 (2004): 34–61.

7. Leonhard Euler, *Elements of Algebra,* ed. J. Hewlett, trans. from the French into English by Francis Horner (London: Longman/Hurst/Rees/Orme, 1822), 5th ed., with a Memoir of the Life and Character of Euler, by F. Horner (London: Longman/Orme, 1840), xlviii.

8. Florian Cajori, *A History of Mathematical Notations* (Chicago: Open Court, Chicago, 1929), 127.

9. The English translation of Euler's *Algebra* was made from the French translation. I give my own, more literal translations of Euler's original German. All subsequent references are thus to the original German edition of 1770, unless otherwise noted.

10. Euler, *Vollständige Anleitung zur Algebra* (1770), Article 131, p. 79 (trans. Martínez).

11. Euler, *Vollständige Anleitung zur Algebra* (1770), Article 329, p. 206 (trans. Martínez).

12. Euler, *Vollständige Anleitung zur Algebra* (1770), Article 206, p. 123 (trans. Martínez): "brauchen wir ein neues Zeichen, welches anstatt der bisher so häufig vorgekommenen Redens-Art, *ist so viel als,* gesetzt werden kann. Dieses Zeichen ist nun = und wird aus gesprochen *ist gleich.*"

13. Euler, *Vollständige Anleitung zur Algebra* (1770), Article 122, p. 72 (trans. Martínez).

14. Euler, *Vollständige Anleitung zur Algebra* (1770), Article 122, pp. 86–88 (trans. Martínez).

15. Leonhard Euler, *Vollständge Anleitung zur Algebra,* ed. J. Niessner, and J. Hofmann (Stuttgart: Reclam-Verlag, 1959), 87.

16. Tristan Needham, *Visual Complex Analysis* (Oxford: Clarendon/Oxford University Press, 1997), 1.

17. I. Grattan-Guinness, *The Norton History of the Mathematical Sciences* (New York: W. W. Norton, 1998), 334.

18. Kline, *Mathematics,* 121; Nahin, *Imaginary Tale,* 12.

19. Euler, *Vollständige Anleitung zur Algebra* (1770), Article 150, p. 88 (trans. Martínez).

20. Euler, "Recherches sur les racines imaginaires des equations," 80–81, 113 (page numbers refer to *Opera Omnia* reprint).

21. Etienne Bézout, *Cours de mathématiques à l'usage du Corps Royal de l'Artillerie, tome 2: Contenant l'algèbre et l'application de l'algèbre a la géométrie* (Paris: P. D. Pierres, Imprimeur ordinaire du Roi, 1781), 95 (trans. Martínez).

22. Bézout, *Cours de mathématiques . . . ,* tome 2, p. 96 (trans. Martínez). See also Etienne Bézout, *Cours de mathématiques, a l'usage des Gardes du Pavillon, de la marine, et des éleves de l'École Politechnique, Troisième partie, contenant l'algebre et l'application*

de cette science à l'arithmétique et la géométrie, nouvelle edition, reviewed and corrected by J. Garnier (Paris: Courcier, Imprimeur Libraire pour les Mathématiques, 1802), 132; Etienne Bézout, *Cours de mathématiques a l'usage de la marine et de l'artillerie, Troisième partie, Contenant l'algèbre et l'application de l'algèbre a la géométrie,* with explanatory notes by A. Reynaud (Paris: Courcier, 1812), 98.

23. Lacroix, *Élémens d'algèbre,* Article 164, pp. 233, 239–40 (trans. Martínez).

24. Jeremiah Day, *An Introduction to Algebra, Being the First Part of a Course of Mathematics, Adapted to the Method of Instruction in American Colleges,* 36th ed. (New Haven: Durrie & Peck, 1839), 324–25.

25. Jean G. Garnier, "Notes et additions à *l'Algèbre* d'Euler," in Leonhard Euler, *Élémens d'Algèbre,* tome 1, new ed., revised and augmented (Paris: Courcier, 1807), 498 (trans. Martínez).

26. Euler, *Vollständige Anleitung zur Algebra* (1770), part 2, p. 165; Euler, *Elements of Algebra* (1840), part 1, Article 753, p. 272.

27. The option of using the $\pm$ sign in writing $\sqrt{-4} = \pm 2i$ is arbitrary; we might just as well write $\sqrt{-4} = \mp 2i$, especially if we wish to write $\sqrt{-4}\,\sqrt{-9} = \mp 2i \times \mp 3i = \mp 6\,(i^2) = \mp 6\,(-1) = \pm 6$.

28. Cajori, *History of Mathematical Notations,* vol. 2, p. 127; Grattan-Guinness, *Norton History of the Mathematical Sciences,* 335.

29. Garnier, "Notes et additions," 491.

30. Isaac Todhunter, *Algebra, for the Use of Colleges and Schools,* 7th ed. (London: MacMillan, 1875), 213–14; Charles Smith, *A Treatise on Algebra,* 2nd ed. (London: MacMillan, 1890), 221.

31. C. Smith, *Treatise on Algebra,* 206.

32. Peacock, *On Symbolical Algebra,* vol. 2 of *A Treatise on Algebra* (Cambridge: Cambridge University Press, 1845), 236.

33. William R. Hamilton, *Theory of Conjugate Functions, or Algebraic Couples; with a Preliminary and Elementary Essay on Algebra as the Science of Pure Time,* based on a lecture of 1833 (Dublin: Philip Dixon Hardy, 1835), reprinted as "Theory of Conjugate Functions," in *Transactions of the Royal Irish Academy* 17 (1837): 293–422.

34. Peacock, *On Symbolical Algebra,* 74–76.

8. The Four of Pythagoras

1. William Wordsworth, *The Excursion, being a portion of The Recluse, A Poem* (London: Longman, Hurst, Rees, Orme, and Brown, 1814), Book 1, pp. 15–16. The words "charm severe" had been used previously in a poem by Anthony Champion, which mentioned "Urania, Muse of charm severe, With chords, that thrill thro' many a sphere, Strings her celestial lyre." See Anthony Champion, "The Empire of Love, A Philosophical Poem" (1770), in *Miscellanies, in Verse and Prose, English and Latin,* ed. William Henry Lord Lyttelton (London: T. Bensley Bolt-Court, 1801), 116.

2. Wordsworth, *Excursion,* Book 4, 197.

3. The discussion between Wordsworth and Hamilton was reported by Hamilton's sister. See Eliza Mary Hamilton, "Wordsworth at the Observatory, Dunsink" (Aug. 1829), in Robert Perceval Graves, *Life of Sir William Rowan Hamilton, including selections from his poems, correspondence, and miscellaneous writings* (Dublin: Hodges, Figgis, & Co., 1882), vol. 1, p. 314.

4. Graves, *Life of Sir William Rowan Hamilton,* vol. 1, p. 315.

5. Hamilton, "Introductory Lecture on Astronomy" (1832), in Graves, *Life of Sir William Rowan Hamilton,* vol. 1, pp. 643–44.

6. Hamilton, *Theory of Conjugate Functions,* 4.

7. W. R. Hamilton, Quaternions, notebook 24.5, entry for 16 Oct. 1843, in *Mathematical Papers of Sir William Rowan Hamilton* (Cambridge: University Press, 1967), vol. 3, p. 103.

8. W. R. Hamilton to his son Archibald H. Hamilton, 5 Aug. 1865, in Graves, *Life of Sir William Rowan Hamilton,* vol. 2, p. 434.

9. W. R. Hamilton to Archibald H. Hamilton, 5 Aug. 1865.

10. W. R. Hamilton to P. G. Tait, 15 Oct. 1858, in Graves, *Life of Sir William Rowan Hamilton,* vol. 2, pp. 435–36.

11. W.R. Hamilton, notebook entry, 16 Oct. 1843, in Hamilton Manuscripts, Library of Trinity College, Dublin, in Graves, *Life of Sir William Rowan Hamilton,* vol. 2, p. 436.

12. W. R. Hamilton to Archibald H. Hamilton, 5 Aug. 1865, in Graves, *Life of Sir William Rowan Hamilton,* vol. 2, pp. 434–35.

13. W. R. Hamilton, notebook entry, 16 Oct. 1843, in Graves, *Life of Sir William Rowan Hamilton,* vol 2, pp. 439–40.

14. W. H. Hamilton to R. P. Graves, 17 Oct. 1843, in Graves, *Life of Sir William Riwab Hamilton,* vol. 2, pp. 441–42.

15. John T. Graves to W. Hamilton, undated letter (1843), in Graves, *Life of Sir William Rowan Hamilton,* vol. 2, p. 443.

16. Graves later learned that the "eight squares theorem" had already been established by C. F. Degen in 1818; see Leonard E. Dickson, "On Quaternions and their Generalization and the History of the Eight Square Theorem," *Annals of Mathematics* 20 (1919): 155–71.

17. Arthur Cayley, "On Jacobi's Elliptic Functions, in reply to the Rev. B. Bronwin; and on Quaternions," *Philosophical Magazine* 26 (1845): 208–11.

18. William Rowan Hamilton, "Recollections of Collingwood. II. The Tetractys" (Observatory, Oct. 1846), in Graves, *Life of Sir William Rowan Hamilton,* vol. 2, p. 525.

19. [Théōnos Smyrnaíou platōnikoû tōn katà tò mathēmatikòn chrēsímōn eis tèn Plátōnos anágnōsiv], in Théon de Smyrne, *Exposition des Connaissances Mathématiques Utiles pour la Lecture de Platon,* trans. J. Dupuis (Paris: Libraire Hachette, 1892), sec. 38, p. 155.

20. *The Life of Pythagoras, with his Symbols and Golden Verses. Together with the Life of Hierocles, and his Commentaries upon the Verses,* trans. into French by M. Darcier, with "The Golden Verses" trans. from Greek into English by Nicholas Rowe (London: Jacob Tonson, 1707), 311; Iamblichus, *On the Pythagorean Way,* chap. 29; Porphyry, *Life of Pythagoras,* sec. 20, p. 127: "I call to witness him who to our souls expressed the Tetraktys, eternal Nature's fountain-spring." The date of "The Golden Verses" is unknown; many scholars trace them roughly to the fifth century CE, although the verses include some lines, such as the oath in question, that are clearly older.

21. [Falsely attributed to Plutarch], *Placita Philosophorum* i.3 [actually by another writer, based on a work by Aetius, ca. 50 BCE, as noted by Theodoret], *Peri tōn areskontōn philosophois physikōn dogmatōn* [and falsely attributed to Qustā ibn Lūqā by Ibn al-Nadīm], in *Aetius Arabus: Die Vorsokratiker in Arabischer Überlieferung,* ed. Hans Daiber (Wiesbaden: Franz Steiner Verlag, 1980), also in Diels, *Doxographi Graeci,* 280; Fairbanks, *First Philosophers of Greece,* 144.

22. Lucian, "Slip of the Tongue," 37; Lucian added, "Philolaus might be quoted."

23. Lucian, "Philopatris; or, the Learner," in *Dialogues of Lucian, from the Greek* (London: N. Longman, 1798), vol. 5, p. 769.

24. W. R. Hamilton to Augustus De Morgan, 24 Dec. 1851, in Graves, *Life of Sir William Rowan Hamilton,* vol. 3, p. 303.

25. W. R. Hamilton to Augustus De Morgan, 14 Feb. 1852, in Graves, *Life of Sir William Rowan Hamilton,* vol. 2, p. 340.

26. Anonymous, review of *Lectures on Quaternions, North American Review* 85, no. 176 (1857): 226, 233.

27. Alexander Macfarlane, review of *Utility of Quaternions in Physics,* by A. McAulay, *Physical Review* 1, no. 5 (1894): 387.

28. Alexander McAulay, *Utility of Quaternions in Physics* (London: Macmillan and Co., 1893), 1.

29. Oliver Heaviside, "The General Solution of Maxwell's Electromagnetic Equations in a Homogenoeus Isotropic Medium, Especially in Regard to the Derivation of Special Solutions, and the Formulae for Plane Waves," *Philosophical Magazine* 27, no. 164 (1889): 45, 48.

30. Oliver Heaviside, "Electromagnetic Theory—XIX," *Electrician* 28, no. 704 (891): 28.

31. Heaviside, "Electromagnetic Theory—XIX," 28.

32. Oliver Heaviside, *Electromagnetic Theory* (London: E. Benn, 1925), vol. 3, p. 135.

33. Oliver Heaviside, "Vectors *versus* Quaternions," *Nature* 47, no. 1223 (1893): 534.

34. Edwin E. Slosson, "Willard Gibbs, Physicist, 1839-1903," in *Leading American Men of Science,* ed. David Starr Jordan (New York: Henry Holt and Company, 1910), 357.

35. Slosson, "Willard Gibbs," 360.

36. P. G. Tait, *An Elementary Treatise on Quaternions,* 3rd ed. (Cambridge: University Press, 1890), vi.

37. Heaviside, "Vectors *versus* Quaternions," 534.

38. Heaviside to Oliver Lodge, 10 Dec. 1908, quoted in Bruce Hunt, *The Maxwellians* (Ithaca: Cornell University Press, 1991), 241.

39. Gottlob Frege, *Die Grundlagen der Arithmetic* (1884), trans. by J. L. Austin as *The Foundations of Arithmetic,* 2nd. ed., rev. (Evanston: Northwestern University Press, 1996), sec. 97, p. 108.

40. Nahin, *Imaginary Tale,* 226.

41. Arnold Dresden, *An Invitation to Mathematics* (New York: Henry Holt and Company, 1936), 35, 83–86.

42. Edna E. Kramer, *The Main Stream of Mathematics* (New York: Oxford University Press, 1951), 113.

43. John C. Baez, "The Octonions," *Bulletin of the Amerian Mathematical Society, New Series* 39, no. 2 (2002): 145–205, quotation at 145.

44. Bell, *Men of Mathematics,* 360–61.

9. The War over the Infinitely Small

1. Emil Wiechert, "Die Theorie der Elektrodynamik und die Röntgen'sche Entdeckung," *Schriften der Physikalisch-Ökonomischen Gesellschaft zu Königsberg in [Preussen]* 37 (Apr. 1896): 1–48, quotation at 3 (trans. Martínez).

2. Frederick Soddy, "Some Recent Advances in Radioactivity," *Contemporary Review* 83 (May 1903): 708–20, quotation at 715.

3. Euler, *Vollständige Anleitung zur Algebra* (1770), trans. by John Farrar as *An Introduction to the Elements of Algebra,* 2nd ed. (Cambridge, MA: University Press, 1821), Article 480, p. 155.

4. Alberto A. Martínez, *Science Secrets: The Truth about Darwin's Finches, Einstein's Wife, and Other Myths* (Pittsburgh: University of Pittsburgh Press, 2011), 1–12; Lane Cooper, *Aristotle, Galileo, and the Tower of Pisa* (Ithaca: Cornell University Press, 1935).

5. For Newton on infinitesimals, see Gert Schubring, *Conflicts between Generalization, Rigor, and Intuition* (New York: Springer, 2005), 161–67.

6. Isaac Newton, *Opera quae Exstant Omnia,* ed. Samuel Horsley (London, 1779), vol. 1, p. 338.

7. Isaac Newton, *Mathematical Principles of Natural Philosophy,* 3rd ed., 1726, trans. Andrew Motte, rev. Florian Cajori (Berkeley: University of California Press, 1934), Book 1, Scholium, 38.

8. Newton, *Mathematical Principles,* Book 1, Scholium, 38–39.

9. John Conduitt, King's College, Cambridge, Keynes Ms. 130.5 (n.d.): "Miscellanea," no. 2, in The Newton Project (Archive), University of Sussex.

10. Gottfried Leibniz to Wilhelm Bierling, 7 July 1711, in *Die Philosophischen Schriften von G. W. Leibniz,* ed. C. I. Gerhardt (Berlin, 1890), vol. 7, p. 497 (trans. Martínez).

11. Leibniz to Pierre Varignon, 2 Feb. 1702, 20 June 1702, both in Gottfried Wilhem Leibniz, *Briefwechsel zwischen Leibniz, Wallis, Varignon, Guido Grandi, Zendrini, Hermann und Freiherrn von Tschirnhaus,* vol. 4 of *Mathematische Schriften,* ed. C. I. Gerhardt (Halle, 1859), 91–92, 110.

12. Leibniz to Johann Bernoulli, Jan. 1699, 21 Feb. 1699, both in *Briefwechsel zwischen Leibniz, Jacob Bernoulli, Johann Bernoulli und Nicolaus Bernoulli,* vol. 3 of Gottfried Wilhem Leibniz, *Mathematische Schriften,* ed. C. I. Gerhardt, vol. 3 (Halle, 1856), part 2, pp. 536, 575.

13. Leibniz to Bernard Nieuwentijt (1695), in Schubring, *Conflicts between Generalization,* 169–71.

14. On Pythagoras and the "monad," see Hippolytus [traditionally misattributed to Origen], Κατα ποσων αιρεσεων ελεγχοσ [*Philosophumena: Refutatio Omnium Haeresium*] (ca. 225 CE), trans. by J. MacMahon as *The Refutation of all Heresies,* ed. Alexander Roberts and James Donaldson (Edinburgh: T&T Clark, 1867), book 4, chap. 15, book 6, chap. 18; anonymous "Commentaries of Pythagoras," quoted by Alexander Polyhistor in Diogenes Laertius, "Life of Pythagoras," sec. 19. See also Iordani Bruni Nolani [Giordano Bruno], *De Monade Numero et Figura Liber consequens Quinque de Minimo Magno & Mensura: item de Innumerabilibus, Immenso, & Infigurabili, seu, De Vniuerso & Mundis libri octo* (Frankfurt: Ioan. VVechelum & Petrum Fischerum, 1591). For the claimed conceptual lineage between Pythagoras's alleged monad and Leibniz, see Gottfried Wilhelm Leibniz, *The Monadology and Other Philosophical Writings,* trans. Robert Latta (Oxford: Clarendon Press, 1898), 34; Ludwig Stein, *Leibniz und Spinoza: Ein Beitrag zur Entwicklungsgeschichte der Leibnizischen Philosophie* (Berlin: Georg Reimer, 1890), 200.

15. Johann Bernoulli, *Oratio Inauguralis in Laudem Mathesos* (1695), quoted in Gerard Sierksma and Wybe Sierksma, "The Great Leap to the Infinitely Small. Johann Bernoulli: Mathematician and Philosopher," *Annals of Science* 56 (1999): 433–39, quotation at 436.

16. Johann Bernoulli, *Die Differentialrechnung aus dem Jahre* (1691–92) (manuscript) (Leipzig: Akademische Verlagsgesellschaft, 1924), 11, trans. in Schubring, *Conflicts between Generalization,* 187.

17. Guillaume-François-Antoine L'Hospital, *Analyse des infiniment petits* (Paris: Imprimerie Royale, 1696), 2–3.

18. Leibniz to François Dangicourt, 1716, quoted in Paolo Mancosu, *Philosophy of Mathematics and Mathematical Practice in the Seventeenth Century* (New York: Oxford University Press, 1996), 72. The secrecy worked, as centuries later Russell

commented: "Leibniz believed in actual infinitesimals, but although this belief suited his metaphysics it had no sound basis in mathematics." See Bertrand Russell, "The Philosophy of Logical Analysis," in *The Basic Writings of Bertrand Russell,* ed. Robert Edward Egner, Lester Eugene Dennon, John Slater (1961; reprint, London: Routledge, 1992), 301.

19. Leibniz to Varignon, 20 June 1702, translation in Schubring, *Conflicts between Generalization,* 173. See also Colin Maclaurin, *A Treatise on Fluxions* (Edinburgh, 1742), 1: 39–40, which notes that Leibniz "owns them [infinites and infinitesimals] to be no more than fictions."

20. Carl B. Boyer, *The History of the Calculus and Its Conceptual Development* (1949; reprint, New York: Dover Publications, 1959), 21.

21. Plato, Πολιτεία (ca. 375 BCE), trans. by Benjamin Jowett as *The Republic of Plato* (London: Oxford University Press, 1881), book 7, p. 221.

22. Euclid, *Elementa* (ca. 250 BCE), book 5, definition 4: "Magnitudes are said to have a ratio to one another, which can, when multiplied, exceed one another." Yet in one instance (book 1, definition 8), the author admits quantities that do not meet that requirement, to describe angles that have boundary lines formed by curves, e.g., hornlike angles.

23. Archimedes, *Method.*

24. George Berkeley, *The Analyst; or, a Discourse Addressed to an Infidel Mathematician* (Dublin: S. Fuller; London: J. Tonson, 1734), pp. 4, 29.

25. Berkeley, *Analyst,* 59.

26. George Berkeley, *A Defence of Free-Thinking in Mathematics. In answer to a pamphlet of Philalethes Cantabrigiensis, intituled, Geometry No Friend to Infidelity* (Dublin: M. Rahmes for R. Gunne; London: J. Tonson, 1735), sec. 21, p. 7.

27. Marcus Tullius Cicero, *De Natvra Deorvm, Book I,* ed. Andrew R. Dyck (Cambridge: Cambridge University Press, 2003), 25, paragraph 5 (trans. Martínez).

28. Newton, *Mathematical Principles,* Book 1, Scholium to Lemma XI, 39.

29. Augustin-Louis Cauchy, *Cours d'analyse* (Paris, 1821), in *Oeuvres complètes d'Augustin Cauchy,* ser. 2, vol. 3 (Paris: Gauthier-Villars, 1899); Augustin-Louis Cauchy, *Résumé des leçons données a l'École Royale Polytechnique sur le calcul infinitésimal,* in *Oeuvres,* ser. 2, vol. 4 (1899).

30. For example, see Augustin-Louis Cauchy, *Sept leçons de physique générale* (Paris: Gauthier-Villars, 1868), 42.

31. Cauchy, *Sept leçons,* 27–28 (trans. Martínez).

32. Cauchy, *Cours d'analyse,* 19.

33. Georg Cantor to Giulio Vivanti, 13 Dec. 1893, in Herbert Meschkowski, "Aus den Briefbüchern Georg Cantor," *Archive for History of Exact Sciences* 2, no. 6 (1965): 505.

34. Cantor to G. Veronese, 7 Sept. 1890, in *Georg Cantor Briefe,* ed. Herbert Meschkowski and Winfried Nilson (Berlin: Springer-Verlag, 1991), 330 (trans. Martínez).

35. Boyer, *History of the Calculus,* 11, 309.

36. "In the nineteenth century infinitesimals were driven out of mathematics once and for all, or so it seemed." This claim appeared first in Martin Davis and Reuben Hersh, "Nonstandard Analysis," *Scientific American* (June 1972): 78–84, reprinted in Davis and Hersh, *Mathematical Experience,* 237–54, quotation at 237–38.

37. Philip Ehrlich, "The Rise of Non-Archimedean Mathematics and the Roots of a Misconception I: The Emergence of Non-Archimedean Systems of Magnitudes," *Archive for History of Exact Sciences* 60 (2006): 1–121.

38. Abraham Robinson, "Non-Standard Analysis," *Proceedings of the Royal Academy of Sciences, Amsterdam,* series A, 64 (1961): 432–40.

39. In addition to the references cited thus far, the contents of table 10 stem partly from the following sources: Euler, *Institutiones Calculi Differentialis* (1755), in *Opera Omnia,* series 1, vol. 10, p. 69; Paul du Bois-Reymond, *Die allgemeine Functionentheorie. Erster Teil: Metaphysik und Theorie der mathematischen Grundbegriffe: Größe, Grenze, Argument und Function* (Tübingen: H. Laupp'schen Buchhandlung, 1882), 71–72; W. Killing, *Die Nicht-Euklidischen Raumformen in Analytischer Behandlung* (Leipzig: B. G. Teubner, 1885), 46–47; Giuseppe Veronese, "Il Continuo Rettilineo e l'Assioma V di Archimede," *Memorie della Reale Accademia dei Lincei, Atti della Classe di Scienze Naturali, Fisiche e Matematiche* 4, no. 6 (1889): 603–24; Giuseppe Veronese, *Fondamenti di Geometria a più Dimensioni e a più Specie di Unità Rettilinee esposti in forma Elementare, Lezioni per la Scuola di Magistero in Matematica* (Padova: Tipografia del Seminario, 1891); Giuseppe Veronese, "Intorno ad alcune Osservazioni sui Segmenti Infiniti o Infinitesimi Attuali," *Mathematische Annalen* 47 (1896): 423–32; Giulio Vivanti, "Sull'Infinitesimo Attuale," *Rivista di Matematica* (Torino) 1 (1891): 135–53, esp. 140; Giuseppe Peano, "Dimostrazione dell'impossibilità di segmenti infinitesimi costanti," *Rivista di Matematica* 2 (1892): 58–62; David Hilbert, *Grundlagen der Geometrie,* Festschrift zur Feier der Enthüllung des Gauss-Weber Denkmals in Göttingen (Leipzig: Teubner, 1899); Bertrand Russell, "Recent Work on the Principles of Mathematics," *International Monthly* 4 (1901): 83–101.

40. Abraham Robinson, "From a Formalist's Point of View," *Dialectica* 23, no. 1 (1969): 45–49.

41. Bertrand Russell, *A History of Western Philosophy* (1945; London: Routledge, 2004), 42–43.

42. Bertrand Russell, *My Philosophical Development* (1959; 2nd ed., London: Routledge, 1995), 154.

43. Timothy Gowers, *Mathematics* (Oxford: Oxford University Press, 2002), 60.

44. Timothy Gowers, "What Is So Wrong with Thinking of Real Numbers as Infinite Decimals?" Department of Pure Mathematics and Mathematical Statistics, University of Cambridge, http://www.dpmms.cam.ac.uk/~wtg10/decimals.html.

10. Impossible Triangles

1. Albert Einstein, "Autobiographisches," trans. by Paul Arthur Schilpp as "Autobiographical Notes," in *Albert Einstein: Philosopher-Scientist,* ed. Paul Arthur Schilpp (Evanston: George Banta Publishing Company, 1949), 9.

2. Plato, *Republic* (Jowett), book 7, p. 222.

3. The translation is from Thomas Heath, *From Thales to Euclid,* vol. 1 of *A History of Greek Mathematics* (Oxford: Clarendon Press, 1921), 141; see also Proclus, *Commentary,* 84.

4. Hippolytus, *Philosophumena,* book 6, section 24.

5. Kepler, *Harmony of the World,* 304.

6. Hamilton, *Theory of Conjugate Functions,* 4.

7. Jean d'Alembert, *Mélanges de littérature, d'histoire, et de philosophie* (1759; new ed., Amsterdam: Zacharie Chatelain & Fils, 1767), vol. 5, sec. 11, p. 207.

8. William Ludlam, *Rudiments of Mathematics* (London: J. & J. Merrill, T. Cadell, B. White, and G. & T. Wilkie, 1785).

9. John Playfair, *Elements of Geometry, containing the First Six Books of Euclid* (Edinburgh: Bell & Bradfute, 1795), 7; John Playfair, *Elements of Geometry,* 2nd. ed. (Edinburgh: Bell & Bradfute, 1804), 402.

10. Gauss, excerpt of 27 Apr. 1813, in Carl Friedrich Gauss, *Werke* (Leipzig: B. G. Teubner, 1900), vol. 8, p. 166 (trans. Martínez).

11. Gauss to Bessel, 27 Jan. 1829, in Gauss, *Werke,* vol. 8, p. 200.

12. F. K Schweikart, Dec. 1818, in Gauss, *Werke,* vol. 8, pp. 180–81; see also pp. 201, 238, 266.

13. George Bruce Halsted, "Bolyai János," *American Mathematical Monthly* 5, no. 2 (1898); 35–37.

14. Farkas Bolyai to Carl Gauss, 20 June 1831, in *Briefwechsel zwischen Carl Friedrich Gauss und Wolfgang Bolyai,* ed. Franz Schmidt and Paul Stäckel (Leipzig: B. G. Teubner, 1899), doc. 33, p. 103.

15. Farkas Bolyai to János Bolyai, 4 Apr. 1820, in Paul Stäckel, *Wolfgang und Johann Bolyai geometrische Untersuchungen, mit Unterstützung der Ungarischen Akademie der Wissenschaften* (Leipzig: B. G. Teubner, 1913), vol. 1, p. 81; translation from *Non-Euclidean Geometries: János Bolyai Memorial Volume,* ed. and trans. András Prékopa and Emil Molnár, vol. 581 of *Mathematics and Its Applications* (New York: Birkhäuser, 2006), 16 (trans. modified).

16. János Bolyai to Farkas Bolyai, 3 Nov. 1823, in Prékopa and Molnár, *Non-Euclidean Geometries,* 16.

17. Nicolai Lobachevsky, *New Principles of Geometry* (1825), trans. and quoted in Roberto Bonola, *Non-Euclidean Geometry: A Critical and Historical Study of its Developments* (1911; reprint, New York: Dover Publications, 1955), 92.

18. [S. Rozhdestvenskii and the Ministerstvo Narodnago Prosveshcheniya], *Istoricheskii obzor deyatel'nosti Ministerstva narodnogo prosveshcheniya 1802-1902* (St. Petersburg: Gos. Tipografiia, 1902), 116, trans. in Hugh Seton-Watson, *The Russian Empire, 1901–1917* (Oxford: Oxford University Press, 1967), 167.

19. Nikolai Bulich, *Ocherki po istorii russkoi literatury i prosvieshcheniia s nachala XIX vieka* (St. Petersburg: Tip. M.M. Stasi'ù`levicha, 1905), vol. 2, p. 271. See also James H. Billington, *Fire in the Minds of Men: Origins of the Revolutionary Faith* (New York: Basic Books, 1980), 141–40.

20. Porphyry, *Life of Pythagoras* (trans. Guthrie), sec. 49, p. 133.

21. For example, Morris Kline, *Mathematical Thought*, vol. 3, pp. 872–73.

22. Ernst Breitenberger, "Gauss's Geodesy and the Axiom of Parallels," *Archive for History of Exact Sciences* 31 (1984): 273–89; Arthur I. Miller, "The Myth of Gauss' Experiment on the Euclidean Nature of Physical Space," *Isis* 63 (1972): 345–48.

23. W. Sartorius von Waltershausen, *Gauss zum Geddchtniss* (1856; reprint, Wiesbaden: Sandig, 1965), 81.

24. Edmund Hoppe, "C. F. Gauss und der Euklidische Raum," *Naturwissenschaften* 13 (1925): 743–44.

25. Jeremy J. Gray, *János Bolyai, Non-Euclidean Geometry, and the Nature of Space* (Cambridge, MA: Burndy Library/MIT Press, 2004), 67–75.

26. Boyai, 1832, quoted in *János Bolyai, Appendix, The Theory of Space*, ed. Ferenc Kárteszi (Amsterdam: Elsevier/Akadémiai Kiadó, 1987), 36.

27. Farkas Bolyai to Gauss, 20 June 1831, in Schmidt and Stäckel, *Briefwechsel zwischen Carl Friedrich Gauss und Wolfgang Bolyai*, doc. 33, p. 103 (trans. Martínez).

28. Franz Schmidt, "Aus dem Leben zweier ungarischer Mathematiker Johann und Wolfgang Bolyai von Bolya," *Archiv der Mathematik und Physik* 48, no. 2 (1868): 227 (trans. Martínez). In the 1890s, George Bruce Halsted translated and edited Bolyai's "The Science of Absolute Space," echoing the story about thirteen duels. See John Bolyai, *The Science Absolute of Space, Independent of the Truth or Falsity of Euclid's Axiom XI*, trans. George Bruce Halsted, 4th ed. (Austin: The Neomon, 1896), xxix.

29. George Bruce Halsted, "Bolyai, Lobatchevsky, Russell," *Monist* 20, no. 1 (1910): 130.

30. George Bruce Halsted, "Bolyai János," *American Mathematical Monthly* 5 (1898): 35.

31. *The Holy Bible, King James Version* (New York: American Bible Society, 1997), vii.

32. Bernhard Riemann, *Über die Hypothesen, welche der Geometrie zu Grunde liegen* (Göttingen: Dietrich, 1867).

33. Eugenio Beltrami, *Saggio di Interpretazione della Geometria Non-Euclidea* (Napoli: De Angelis, 1868).

34. John Earman and Clark Glymour, "Relativity and Eclipses: The British

Eclipse Expeditions of 1919 and their Predecessors," *Historical Studies in the Physical Sciences* 11 (1980): 49–85.

35. John Waller, "The Eclipse of Isaac Newton: Arthur Eddington's 'Proof' of General Relativity," in John Waller, *Einstein's Luck: The Truth behind Some of the Greatest Scientific Discoveries* (Oxford: Oxford University Press, 2002), 48–63; Tony Rothman, "What Did the Eclipse Expedition Really Show? And Other Tales of General Relativity," in *Everything's Relative,* 77–87; "Are the Stars Displaced in the Heavens?" in Harry Collins and Trevor Pinch, *The Golem: What Everybody Should Know about Science* (Cambridge: University Press, 1993), 43–55.

36. Daniel Kennefick, "Not Only because of Theory: Dyson, Eddington and the Competing Myths of the 1919 Eclipse Expedition," in Christoph Lehner, Jürgen Renn, and Matthias Schemmel, eds., *Einstein and the Changing Worldviews of Physics* (New York: Birkhäuser, 2012), 201–32. There were fair scientific grounds on which to discard plates from the Greenwich Observatory's Astrographic lens (at Sobral); it was not arbitrarily discarded because of Eddington's bias in favor of Einstein's theory. Those plates were actually rejected by Frank Dyson (who was skeptical of Einstein's theory), apparently because information about a necessary scale value on that lens was lacking. In 1978, such plates were remeasured at the Royal Greenwich Observatory, yielding results much closer to Einstein's prediction.

37. J. J. Thomson, quoted in "Eclipse Showed Gravity Variation. Dispersion of Light Rays Accepted as Affecting Newton's Principles," *New York Times,* 9 Nov. 1919.

38. "Lights All Askew in the Heavens. Men of Science More or Less Agog over Results of Eclipse Observations. Einstein Theory Triumphs," *New York Times,* 10 Nov. 1919.

39. "Lights All Askew" (the name of the astronomer is not specified).

40. Einstein to Heinrich Zangger, early 1920, in *Collected Papers of Albert Einstein* (Princeton: Princeton University Press, 2004), vol. 9, p. 204.

41. Einstein to Max and Hedwig Born, 9 Sept. 1920, in Albert Einstein and Max Born, *Briefwechsel 1916-1955,* ed. Max Born (Munich: Nymphenburger Verlagshandlung, 1969), 59 (trans. Martínez).

42. Anthony M. Alioto, *A History of Western Science,* 2nd ed. (New Jersey: Prentice Hall, 1993), 373.

43. Francis D. Murnaghan, "The Quest for the Absolute: Modern Developments in Theoretical Physics and the Climax Supplied by Einstein," in *Relativity and Gravitation,* ed. J. Malcolm Bird (London: Methuen, 1921), 281.

44. "Lights All Askew."

45. Plato, *Timaeus* (ca. 360 BCE), trans. Benjamin Jowett, in *The Dialogues of Plato,* 3rd ed. (London: Oxford University Press, 1892).

46. Nicolai Copernici, *De Revolutionibus Orbium Cœlestium, Libri VI* (Norimbergae: Ioh. Petreium, 1543), book 1, chap. 4, folio ii, verso.

47. Copernici, *De Revolutionibus Orbium Cœlestium,* book 1, chap. 1, folio i, reverso.

48. Giordano Bruno, *La Cena de le Cerneri* (1584), part 3, p. 165 (trans. Martínez).

49. In a letter of 21 July 1612, Federico Cesi mentions Kepler's elliptical orbits to Galileo, but thereafter Galileo does not discuss this in his works; see doc. 732 in *Le Opere di Galileo Galilei, Edizione Nationale,* ed. Antonio Favaro (Firenze: G. Barbera, 1901), vol. 11, p. 366.

50. Mario D'Addio, *The Galileo Case: Trial, Science, Truth,* trans. Brian Williams (Rome: Nova Millennium Romae, 2004), 75.

51. Galilei, manuscript: "On Bellarmine's 'Letter to Foscarini'" (1615), trans. in Richard J. Blackwell, *Galileo, Bellarmine, and the Bible: Including a Translation of Foscarini's Letter on the Motion of the Earth* (Notre Dame: University of Notre Dame Press, 1991), 269–76, 269.

52. In *Reply to Ignoli* (Oct. 1624), an extensive critique of an anti-Copernican text of 1616 by Francesco Ingoli, Galileo continued: "I have many confirmations of this proposition, but for the present one alone suffices, which is this. I suppose the parts of the universe to be in the best arrangement, so that none is out of its place, which is to say that Nature and God have perfectly arranged their structure. This being so, it is impossible for those parts to have it from Nature to be moved in straight, or in other than circular motion, because what moves straight changes place, and if it changes place naturally, then it was at first in a place preternatural to it, which goes against the supposition. Therefore, if the parts of the world are well ordered, straight motion is superfluous and not natural, and they can only have it when some body is forcibly removed from its natural place, to which it would then return by a straight line, for thus it appears to us that a part of the earth does [move] when separated from its whole. I said 'it appears to us,' because I am not against thinking that not even for such an effect does Nature make use of straight motion." See Favaro, *Le Opere di Galileo Galilei,* vol. 6, p. 558–59, trans. in Stillman Drake, *Galileo at Work, His Scientific Biography* (Chicago: University of Chicago Press, 1978), 294–95. Galileo reaffirmed this outlook in *Dialogue Concerning the Two Chief World Systems* (1632).

53. Galileo Galilei, *Dialogue Concerning the Two Chief World Systems—Ptolemaic & Copernican,* trans. Stillman Drake (Berkeley: University of California Press, 1953), "Second Day," 259.

54. Kepler, *Harmony of the World,* 186.

55. Kepler, *Harmony of the World,* 146.

56. Christiani Hugenii, Κοσμοθεορος, *sive De Terris Cœlestibus, earumque ornatu, Conjecturæ* (The Hague: Adrianum Metjens, 1698); Christianus Huygens, *The Celestial Worlds Discover'd: or, Conjectures concerning the Inhabitants, Plants and Productions of the Worlds in the Planets,* 2nd ed. (London: James Knapton, 1722), 148.

57. The Standard Model of particle physics requires that the distribution of electric charge on any electron is not perfectly spherical, so for years, physicists have assumed that the shape of an electron is not perfectly spherical. The Standard

Model predicts that the asymmetry in the distribution of charge on electrons is so very tiny that it is virtually undetectable, and thus it lies beyond the reach of present experiments; but such experiments do at least provide evidence against certain theories similar to the Standard Model. In 2011, physicists at Imperial College London published extraordinarily accurate experiments, carried out for years, that did not detect any wobbles in the motions of electrons. They interpret these negative results as evidence that electrons are the roundest objects known. They continue to refine the experiments to increase precision. See J. Hudson, D. Kara, I. Smallman, B. Sauer, M. Tarbutt, E. Hinds, "Improved Measurement of the Shape of the Electron," *Nature* 473 (May 2011): 493–96.

58. Werner Heisenberg, "First Encounter with the Atomic Concept (1919-1920)," in *Physics and Beyond* (New York: Harper & Row, 1971), 2–3.

59. [Falsely attributed to Plutarch], *Placita Philosophorum,* 131 (trans. Martínez).

60. Galileo Galilei, *The Assayer* (1623), reissued in *Discoveries and Opinions of Galileo,* trans. Stillman Drake (New York: Anchor Books/Random House, 1957), 237–38.

61. G. H. Hardy, *A Mathematician's Apology* (New York: Cambridge University Press, 1940), 150.

11. Inventing Mathematics?

1. Betty (ninth-grade student), and Vered (eleventh-grade student), quoted in Gilah C. Leder, Erkki Pehkonen, and Günter Törner, *Beliefs: A Hidden Variable in Mathematics Education?* (New York: Springer, 2003), 338–39.

2. See *The New York Times Biographical Service* (New York: Arno Press, 1977), vol. 8, p. 111 (emphasis added); and *How Many Questions? Essays in Honor of Sidney Morgenbesser,* ed. Leigh Cauman et al. (Indianapolis: Hackett, 1983), 1.

3. Sylvestre François Lacroix, *Essais sur l'Enseignement en Général et sur celui des Mathématiques en Particulier* (Paris: Courcier, Imprimeur-Libraire pour les Mathématiques, 1805), 284 (trans. Martínez).

4. Martínez, *Negative Math,* 110–205.

5. Ludwig Wittgenstein, *Remarks on the Foundations of Mathematics,* ed. Gertrude Elizabeth Margaret Anscombe and Rush Rhees (Cambridge, MA: MIT Press, 1956), 229.

6. For example, see Leonard Eugene Dickson, *Algebras and Their Arithmetics* (Chicago: University of Chicago Press, 1923).

7. For examples, see Martínez, *Negative Math,* 132–86.

8. Georg Cantor, "Ueber unendliche, lineare Punktmannichfaltigkeiten," *Mathematische Annalen* 21 (1883): 563–64 (trans Martínez).

9. Arthur Schoenflies, "Die Krisis in Cantor's Mathematischem Schaffen," *Acta Mathematica,* 50 (1927): 1–23, esp. 2: "Es übersteigt nicht das erlaubte Mass, wenn

ich sage, dass die Kroneckersche Einstellung den Eindruck hervorbringen musste, als sei Cantor in seiner Eigenschaft als Forscher und Lehrer ein Verderber der Jugend."

10. Kronecker, quoted in Dauben, *Georg Cantor,* 134–35; Cantor, letter of 6 Jan. 1884, in Meschkowski and Nilson, *Georg Cantor: Briefe* 163 (trans. Martínez).

11. Cantor to Ignatius Jeiler, 1888, quoted in Dauben, *Georg Cantor,* 147.

12. Cantor to Giulio Vivanti, 13 Dec. 1893, in Meschkowski, "Aus den Brief-büchern Georg Cantor," 505 (trans. Martínez).

13. Cantor to Thomas Esser, 15 Feb. 1896, quoted in Dauben, *Georg Cantor,* 147.

14. Dauben, *Georg Cantor,* 229.

15. Henri Poincaré, "L'avenir des mathématiques," *Rendiconti del Circolo Matematico di Palermo* 26 (1908): 152–68, quotation at 167 (trans. Martínez).

16. Henri Poincaré, "Les mathématiques et la logique," in Henri Poincaré, *Dernières Pensées* (Paris: Ernest Flammarion, 1913), 160–61 (trans. Martínez).

17. Poincaré, "La logique de l'infini," in Poincaré, *Dernières Pensées,* 104–5 (trans. Martínez).

18. Poincaré, "L'avenir des mathématiques," 167–68 (trans. Martínez).

19. Poincaré, "Les mathématiques et la logique," 143 (trans. Martínez).

20. David Hilbert, "Über das Unendliche," *Mathematische Annalen* 95 (1925): 161–90, esp. 170 (trans. Martínez).

21. Ludwig Wittgenstein, *Wittgenstein's Lectures on the Foundations of Mathematics, Cambridge, 1939,* ed. Cora Diamond, from the notes of R. G. Bosanquet, Norman Malcolm, Rush Rhees, and Yorick Smythies (Ithaca: Cornell University Press, 1976), lecture 11, p. 103.

22. Wittgenstein, *Wittgenstein's Lectures,* 140–41.

23. Wittgenstein, *Remarks on the Foundations of Mathematics,* 264–65.

24. Davis and Hersh, "Confessions of a Prep School Math Teacher," 272–74.

25. Davis and Hersh, *Mathematical Experience,* 318.

26. Heinrich M. Weber, "Leopold Kronecker," *Jahresberichte der Deutschen Mathematiker Vereinigung* 2 (1893): 5–31, quotation at 19 (trans. Martínez); see also *Mathematische Annalen* 43 (1893): 1–25.

27. Errett Bishop, *Foundations of Constructive Analysis* (New York: Academic Press, 1967).

28. E. J. Brower, *Brower's Cambridge Lectures on Intuitionism,* ed. D. Van Dalen (New York: Cambridge University Press, 1981).

29. J. Donald Monk, quoted in Davis and Hersh, *Mathematical Experience,* 322. See also J. Donald Monk, "On the Foundations of Set Theory," *American Mathematical Monthly* 77 (1970): 703–11.

30. Davis and Hersh, *Mathematical Experience,* 321. See also Jean Dieudonné, "The Work of Nicholas Bourbaki," *American Mathematical Monthly* 77 (1970):

134–45; P. J. Cohen, "Comments on the Foundation of Set Theory," in *Axiomatic Set Theory: Proceedings of Symposia in Pure Mathematics,* ed. Dana Scott (Providence: American Mathematical Society, 1971), 9–15.

12. The Cult of Pythagoras

1. Singh, *Fermat's Enigma,* 7, 8, 17, 23, 25.

2. Fowler, *Mathematics of Plato's Academy,* 358.

3. Porphyry, *Life of Pythagoras* (trans. Guthrie), sec. 28, p. 128.

4. *Genius: Pythagoras,* written by Vanessa Tovell, narr. Bob Session, dir. Jeremy Freeston, prod. Ruth Wood (Cromwell Productions, 1996). Such claims are still made in recent books as well; e.g., "No one has done more," according to Hakim, *Story of Science,* 81.

5. Hilary Gati, "Giordano Bruno and the Protestant Ethic," in *Giordano Bruno: Philosopher of the Renaissance,* ed. Hilary Gati (Cornwall: Ashgate, 2002), 144–66, quotation at 166.

6. Koestler, *Sleepwalkers,* 37.

7. George Orwell, "Why I Write," in George Orwell, *A Collection of Essays* (London: Harcourt Brace Jovanovich, 1946), 316.

8. Iamblichus, *On the Pythagorean Way,* chap. 18, pp. 111–13.

9. Diogenes Laertius, "Life of Pythagoras," sec. 13.

10. George Crabb, *A Dictionary of General Knowledge; or, an Explanation of Words and Things connected with All the Arts and Sciences* (London: Thomas Tegg, 1830), 174. See also Thomas Taylor, *The Theoretic Arithmetic of the Pythagoreans* (1816), with an introductory essay by Manly Hall (Los Angeles: Phoenix Press, 1934), 38.

11. Singh, *Fermat's Enigma,* 50.

12. Crabb, *Dictionary of General Knowledge,* 38.

13. Diogenes Laertius, "Life of Pythagoras," sec. 25.

14. Koestler, *Sleepwalkers,* 26. (In the original, *word* is mistakenly spelled *world.*)

15. Koestler, *Sleepwalkers,* 42.

16. *Harper's Book of Facts: A Classified History of the World, embracing Science, Literature and Art,* comp. Joseph H. Willsey, ed. Charlton Thomas Lewis (New York: Harper & Brothers, 1895), 270.

17. Gottfried Leibniz to Wilhelm Bierling, 7 July 1711, in *Die Philosophischen Schriften von G. W. Leibniz,* vol. 7, p. 497.

18. Theophilus of Antioch, *Apologia ad Autolycum* (ca. 175? CE), in *The Writings of the Early Christians of the Second Century,* trans. J. A. Giles (London: John Russell Smith, 1857), "His Writings to Autolycus," book 3, sec. 7, p. 172.

19. Isaac Newton, *Mathematical Principles of Natural Philosophy* (1687), trans. Andrew Motte (1729) (New York: Daniel Adee, 1846), book 3, General Scholium (1713),

505; Justinus Martyr, *Cohort ad Græcos* (ca. 150 CE), chap. 19, trans. as "Justin's Hortatory Address to the Greeks," in *Writings of Justin Martyr and Athenagoras,* 305. Some scholars have argued that the attribution to Justinus Martyr (who died in 165 CE) is apocryphal and that this work is likely from sometime between 180 and 240 CE; see Jules Lebreton, "St. Justin Martyr," *The Catholic Encyclopedia* (New York: Robert Appleton Company, 1910), vol. 8.

20. Hermippus of Smyrna (ca. 280? BCE), as quoted by Flavius Josephus in *Contra Apionem* (ca. 97 CE), in *The Works of Flavius Josephus, the Learned and Authentic Jewish Historian and Celebrated Warrior,* vol. 4, trans. William Whiston (1737) (London: Lackington, Allen & Co., 1806), "Against Apion," book 1, sec. 22, p. 300.

21. André Dacier, *La vie de Pythagore, ses symboles, ses vers dorez, & la vie d'Hierocles,* (Paris: Rigaud, 1706), vol. 1, pp. lvj–lvij. See also Albert G. Mackey, *An Encyclopedia of Freemasonry and its Kindred Sciences* (1873), rev. ed. by William J. Hughan and Edward L. Hawkins (New York: The Masonic History Company, 1919), vol. 2, p. 780. These writers further claimed that Pythagoras realized that since God's true name has four letters in Hebrew he then translated it as the "tetractys" (or quaternion), the number that signified "the source of nature that perpetually rolls along," which became venerated by the Pythagoreans.

22. Ralph Cudworth, *The True Intellectual System of the Universe* (London, 1678), 22, 187, 288, 413.

23. Jean-Jacques Barthélemy, "Entretien sur l'Institut de Pythagore," in *Voyage d'Anacharsis en Gréce, vers le Milieu du Quatrième Siècle avant l'Ere Vulgaire* (1789) (Paris: E. A. LeQuien, 1822), vol. 6, chap. 75, pp. 267–68.

24. Evan Powell Meredith, *The Prophet of Nazareth* (London: F. Farrah, 1864), 516.

25. *Harper's Book of Facts,* 639.

26. Sylvain Maréchal, *Dictionnaire des Athées anciens et modernes,* 2nd ed. (Brussels: J. B. Balleroy, 1833), 235–37.

27. Annie Bessant, "Introduction," in *The Golden Verses of Pythagoras and Other Pythagorean Fragments,* ed. Florence M. Firth (Hollywood: Theosophical Publishing House, 1904), ix.

28. For example: "Pythagoras . . . promulgated a theory for the preservation of political power in the educated class, and ennobled a form of government which was generally founded on popular ignorance and on strong class interests. He preached authority and subordination, and dwelt more on duties than on rights, on religion than on policy; and his system perished in the revolution by which oligarchies were swept away." See John Emerich Edward Dalberg-Acton, *The History of Freedom and Other Essays,* ed. John Neville Figgis and Reginald Vere Laurence (London: Macmillan and Co., 1907), 21. Likewise, Winspear argued that the Pythagoreans were conservatives who represented the "antipopular" interests of landed aristocrats, "as defenders of the old, the established, and governing against innovation and

anarchy." See Alban Dewes Winspear, *The Genesis of Plato's Thought,* 2nd ed. (New York: S. A. Russell, 1956), 86.

29. For example, Doyne Dawson, *Cities of the Gods: Communist Utopias in Greek Thought* (Oxford: Oxford University Press, 1992), 14–18. The impression of a resemblance to communism is based partly on accounts by Timaeus of Tauromenium, Diodorus Siculus, Diogenes Laertius, and Iamblichus stating that Pythagorean disciples shared or were required to share their property.

30. Karl Kautsky, *Thomas More und seine Utopie, mit einer historischer Einleitung* (Stuttgart: J. H. W. Dietz, 1888), 1. Kautsky claimed that although some labeled Pythagoras a communist, such tendencies of the Pythagoreans were a later development; see Karl Kautsky, *Von Plato bis zu den Wiedertäufern,* vol. 1, part 1 of *Die Vorläufer des Neueren Sozialismus* (Stuttgart: J. H. W. Dietz, 1895), 6–7.

31. George Thomson, *Æschylus and Athens: A Study in the Social Origins of Drama* (London: Lawrence & Wishart, 1946), 210.

32. Alfred Cook, *Psychology, An Account of the Principal Mental Phenomena* (New York: Hinds, Noble & Eldredge, 1904), 173.

33. Cook, *Psychology,* 174.

34. Benjamin Farrington, *Greek Science; Its Meaning for Us* (Harmondsworth, UK: Penguin Books, 1949), vol. 1, p. 43.

35. The interviews were conducted by Harriet Zuckerman in the early 1960s. Merton did not state the name of the laureate in physics he quoted, and he added the word in brackets. See Robert K. Merton, "The Matthew Effect in Science," *Science,* new series, 159, no. 3810 (1968): 56-63, quotation at 57.

36. Robert K. Merton, *The Sociology of Science,* ed. Norman Strorer (Chicago: University of Chicago Press, 1973), chap. 2; Harriet Zuckerman, "Nobel Laureates: Sociological Studies of Scientific Collaboration" (PhD diss., Columbia University, 1965); Harriet Zuckerman, *Scientific Elite: Nobel Laureates in the United States* (New York: Free Press, 1977); Harriet Zuckerman, "Interviewing an Ultra-Elite," *Public Opinion Quarterly* 36 (1972): 159–75.

37. Robert K. Merton, "The Matthew Effect in Science, II: Cumulative Advantage and the Symbolism of Intellectual Property," *Isis* 79, no. 4 (1988): 606–23, esp. 609; Mark, KJV 4:25: "For he that hath, to him shall be given: and he that hath not, from him shall be taken even that which he hath"; Luke, KJV 8:18: "Take heed therefore how ye hear: for whosoever hath, to him shall be given; and whosoever hath not, from him shall be taken even that which he seemeth to have"; Luke, KJV 19:26: "For I say unto you, That unto every one which hath shall be given; and from him that hath not, even that he hath shall be taken away from him."

38. Merton, in "Matthew Effect in Science, II," quoted a lecture by M. de Jonge on "The Matthew Effect" (24 July 1987), wherein de Jonge specifically cited Proverbs 9:9, Daniel 2:21, and Martialis, epigram 5, p. 81: "Semper pauper eris, si pauper es, Aemiliane. Dantur opes nullis [nunc] nisi divitibus." De Jonge concluded: "The use

made of this sentence [in Matthew] by modern authors neglects the eschatological thrust inherent in the saying in all versions, and (in all probability) in Jesus's own version of it. It links up, however, with the Wisdom-saying taken over by Jesus: 'Look around you and see what happens: If you have something, you get more; if you have not a penny, they will take from you the little you have.'" See Merton, *Sociology of Science,* chap. 2; Zuckerman, "Nobel Laureates"; Zuckerman, *Scientific Elite;* Zuckerman, "Interviewing an Ultra-Elite."

39. Richard Holt, "Apollo 11 Moon Landing: Rocketing from Fantasy to Reality," *Telegraph* (UK), 16 July 2009.

40. Frank Manuel, *A Portrait of Isaac Newton* (Cambridge, MA: Belknap/Harvard University Press, 1968), 29.

41. David Gregory, *The Elements of Astronomy, Physical and Geometrical*, 2 vols. (1715), vol. 1, pp. ix–xi, quotation at xi. This work was first published as *Astronomiae Physicae et Geometricae Elementa* (1703).

42. *Harper's Book of Facts,* 742.

43. Koestler, *Sleepwalkers,* 27.

44. Martínez, *Science Secrets,* chap. 2.

45. Helena Petrovna Blavatsky, *The Secret Doctrine: The Synthesis of Science, Religion, and Philosophy,* 3rd ed. (London: Theosophical Society, 1888), vol. 1, p. 676.

46. For example, see Augustine of Hippo, *De Civitate Dei Contra Paganos* (ca. 412–27 CE); *City of God Against the Pagans,* book 7 (ca. 417 CE), chap. 35.

47. Marie François de Aroüet de Voltaire, "Aventure Indienne [de Pythagore]," in *Le philosophe ignorant,* rev. ed. (Paris: Datiment/Mondhare, 1767), 87.

48. E. Cobham Brewer, *Dictionary of Phrase and Fable: Giving the Derivation, Source, or Origin of Common Phrases, Allusions and Words that have a Tale to Tell,* rev. ed. (London: Cassell and Company, 1900), 1022.

49. Hellen Wells, *Pythagoras Speaks: The Science of Numbers and the Art of Will-Power in Language of Today* (New York: Spiritual and Ethical Society of New York City, 1934), 2.

50. Diogenes Laertius, "Life of Pythagoras," sec. 19.

51. Pliny the Elder, *Naturalis Historia* (ca. 79 CE), book 18, sec. 30.

52. Byrne, *Secret,* 4.

53. Jean-Jacques Barthélemy, "Entretien sur l'Institut de Pythagore," in *Voyage du Jeune Anacharsis en Grèce, dans le Milieu du Quatrième Siècle avant l'Ère Vulgaire,* 1st ed. (Paris: De Bure, de la Bibliothèque du Roi, 1789), vol. 8, chap. 75, p. 125 (trans. Martínez).

54. Barthélemy, "Entretien sur l'Institut de Pythagore" (1789), 129 (trans. Martínez).

55. Barthélemy, "Entretien sur l'Institut de Pythagore" (1789), 113 (trans. Martínez).

56. Sylvain Maréchal, *Voyages de Pythagore en Égypte, dans la Chaldée, dans l'Inde,*

en Crète, à Sparte, en Sicile, à Rome, à Carthage, à Marseille et dans les Gaules: suivis de ses Lois Politiques et Morales (Paris: Deterville, 1799).

57. Yu. Oksman, "'Pifagorovy zakony' i 'Pravila soedinennykh slavian,'" in *Ocherki iz istorii dvizhenii'a dekabristov: sbornik statei,* ed. Nikolai Druzhinin and Boris Syroechkovskogo, Akademia nauk SSSR Institut istorii (Moscow: Gos. izd-vo polit. lit-ry, 1954), 485–87, 490. The "Rules" appeared in Maréchal, *Voyages de Pythagore,* vol. 6.

58. See Billington, *Fire in the Minds of Men,* 104–5.

59. Druzhinin and Syroechkovskogo, *Ocherki iz istorii dvizhenii'a dekabristov,* 477–78.

60. Lenin, "Filosofskie tetradi" (1923), 29: 223–25, cited in Frank Ellis, "Soviet Russia through the Lens of Classical Antiquity: An Analysis of Greco-Roman Allusions and Thought in the Oeuvre of Vasilii Grossman," in *Russian Literature and the Classics,* ed. Peter Barta, David Larmour, and Paul Allen (Amsterdam: Overseas Publishers Association, 1996), 138.

61. Herodotus, *The Histories* (ca. 430 BCE), trans. A. D. Godley (Cambridge, MA: Harvard University Press, 1920), sec. 95.

62. Marcus Tullius Cicero, *In Vatinium Testem* (56 BCE), in *The Speech of M. T. Cicero against Publius Vatinius: Called Also, the Examination of Publius Vatinius,* ed. C. D. Yonge (London: George Bell & Sons, 1891), sec. 14.

63. Diogenes Laertius, "Life of Pythagoras," sec. 4.

64. Russell, *History of Western Philosophy,* 38.

65. Russell, *History of Western Philosophy,* 38, 39, 45.

66. Bertrand Russell, "How to Read and Understand History," in *Understanding History, and Other Essays* (New York: Philosophical Library, 1957), 42.

ILLUSTRATION SOURCES AND CREDITS

Figures 2.1, 2.2, 2.3, 3.1, 3.2, 3.3, 3.4, 3.5, 3.6, 3.7, 3.8, 3.9, 6.2, 8.1, 9.1, 9.2, 10.1, 10.2, 10.3, 10.4, 10.5, 10.6, 10.7, 10.8, 10.10, 10.11, 10.12, 10.13, 10.14, 10.15, 10.16, 10.17, 10.18: © Alberto A. Martínez.

Figure 1.1: Euclid, *Preclarissimus Liber Elementorum Euclidis Perspicassimi in Artem Geometrie* (Venetiis, 1482), proposition 47.

Figure 2.4: Johannes Kepler, *Harmonices Mundi Libri V* (Lincii, Austria: Godofredi Tampachii, 1619), 180.

Figures 5.1, 5.2, 5.3: Adolf Zeising, *Neue Lehre von den Proportionen des Menschlichen Körpers, aus einem bisher unerkannt gebliebenen, die ganze Natur und Kunst durchdringenden morphologischen Grundgesetze entwickelt und mit einer vollständigen historischen Uebersicht der bisherigen Systeme* (Leipzig: Rudolph Weigel, 1854), 354–57, 178, 235.

Figure 6.1. Leonhard Euler, *Elements of Algebra*, translated by Peter Barlow from the French, with the additions of Joseph-Louis La Grange, 2nd ed. (London: J. Johnson and Co., 1810), vol. 1, chap. 7, p. 34.

Figure 10.4. *Is God a Geometer?* Image by Alberto A. Martínez, based on the frontispiece of Hadrianus Junius, *Hadriani Ivnii Medici Emblemata* (Antwerp, 1565).

Figure 10.19. Lambsprinck, *De Lapide Philosophico* (Francofurti: Lucæ Jennis, 1625), p. 11, in *Musæum Hermeticum, Omnes Sopho-Spagyricae Artis Discipulos Fidelissime Erudiens, quo Pacto Summa illa veraque medicina . . .* (Francoforti: Lucæ Jennis, 1625).

INDEX